STUDIES IN HISTORY, ECONOMICS AND
PUBLIC LAW

Edited by the
FACULTY OF POLITICAL SCIENCE
OF COLUMBIA UNIVERSITY

NUMBER 472

THE PRODUCTIVITY OF LABOR IN THE
RUBBER TIRE MANUFACTURING INDUSTRY

BY

JOHN DEAN GAFFEY

THE
PRODUCTIVITY OF LABOR
IN THE
RUBBER TIRE
MANUFACTURING
INDUSTRY

BY

JOHN DEAN GAFFEY, Ph. D.

INSTRUCTOR IN ECONOMICS AND BUSINESS ADMINISTRATION
RENSSELAER POLYTECHNIC INSTITUTE

NEW YORK
COLUMBIA UNIVERSITY PRESS
LONDON: P. S. KING & SON, LTD.
1940

COPYRIGHT, 1940

BY

COLUMBIA UNIVERSITY PRESS

PRINTED IN THE UNITED STATES OF AMERICA

to

A. E. G.

PREFACE

THIS study was undertaken in order to present the technological and economic history of the rubber tire manufacturing industry from the standpoint of productivity. The movements of productivity have been described with a view toward facilitating the analysis of the underlying conditions which lead to industrial progress. An attempt also has been made to determine and to interpret the incidence of productivity gains.

My interest in the problems of productivity dates back to the summer of 1936, when I was employed as a field agent in the Productivity Survey Division of the joint studies of the National Bureau of Economic Research and the National Research Project of the Works Progress Administration. My work at that time was under the direction of Mr. David Weintraub and under the more immediate supervision of Dr. William A. Neiswanger and the late Dr. Harry Jerome. My interest in the field was further stimulated by Professor Leo Wolman of Columbia University, under whose guidance the present study was made. Professor Wolman's criticisms, comments, and advice have been invaluable throughout the preparation of the work. I am indebted to Professors Frederick C. Mills, John Maurice Clark, and Paul F. Brissenden, all of Columbia University, who have read the manuscript and offered many valuable suggestions. My appreciation is also extended to Dr. Arthur F. Burns of the National Bureau of Economic Research and to my colleague, Mr. Frank J. Kottke of Rensselaer Polytechnic Institute, for their many helpful comments.

The materials for the study were for the most part gathered during 1937-38, while I was a Pre-Doctoral Field Fellow of the Social Science Research Council. I am indebted to the Council not only for generous financial aid, but also for assistance in gaining access to much of the data required. I also wish to thank the following persons and organizations for courteous aid in making material available to me: Dr. A. F. Hinrichs, Dr. N. A. Tolles, and Mr. Boris Stern, all of the United States Department of Labor, Mr. E. G. Holt of the United States Department of Commerce, Mr. C. W. Halligan of the Rubber Manufacturers Association, Messrs. Sherman H. Dalrymple and Thomas F. Burns, of the United Rubber Workers of America, Mr. W. F. Bloor of the Goodyear Tire and Rubber Company, Mr. Leonard Smith of the United States Rubber Company, Mr. P. L. Dildine of the B. F. Goodrich Company, Mr. J. C. Roberts of the Firestone Tire and Rubber Company, Mr. C. F. Burke of the General Tire and Rubber Company, Mr. F. B. Immler of the Seiberling Rubber Company, Mr. G. E. Sauvain of the Mohawk Rubber Company, Mr. Carl Pharis of the Pharis Tire and Rubber Company, and Mr. John B. Lewis of *Tire Review*.

J. D. G.

TROY, NEW YORK,
MARCH, 1940.

CONTENTS

                                                                       PAGE

CHAPTER I
INTRODUCTION

The Problem.. 15
Previous Investigations in the Field............................... 16
Methods and Procedures... 21

CHAPTER II
THE ORIGINS AND ECONOMIC CHARACTERISTICS OF THE TIRE INDUSTRY

Emergence of the Tire Industry..................................... 25
The Economic Characteristics of the Tire Industry................. 30
    A. Youth... 30
    B. Tire Demand... 31
    C. Tire Supply... 32

CHAPTER III
PRODUCTION

Tires Produced and Rubber Consumed................................ 35
Improvement in Quality.. 37
Tire-Miles Produced... 40
Factors Responsible for the Increase in Tire Durability........... 41
    A. General Forces.. 41
    B. Changes in Type of Tires Produced........................... 42
    C. Developments in Rubber Chemistry............................ 47
    D. Other Factors... 49
Sales and Markets... 50
Physical Productive Capacity...................................... 58

CHAPTER IV
THE LABOR SUPPLY

Employment and Man-Hours.. 62
Characteristics of the Labor Force................................ 64

CHAPTER V
THE PRODUCTIVITY OF LABOR

Total Manufacturing Labor Productivity—Direct and Indirect........ 68
Direct Labor Productivity... 72
Direct Labor Productivity by Departments of Tire Manufacturing.... 74
Productivity in Value Terms....................................... 77
The Physical Productivity of the Tire Industry Compared with the Rest of the Economy... 78

9

CHAPTER VI
Factors Conditioning the Increase in Productivity

General Factors	84
Mechanical Factors	89
Non-Mechanical Factors	94
A. Developments in Industrial Management	94
1. Task Standardization	94
2. Incentive Wage Systems	95
3. Effects upon Productivity	96
B. Shortening of the Work Day and Work Week	97
C. The Skill and Efficiency of the Workers	103
The Influence of Labor Relations on Productivity	106
Cyclical and Seasonal Influences on Productivity	118

CHAPTER VII
Some Effects of Increasing Productivity

Effects upon Employment	122
The Distribution of the Gains of Productivity Increase	125
A. Labor Costs	129
B. Gains to Consumers	130
C. Gains to the Owners and Managers	131
1. The Owners	131
2. The Managers	136
3. Other Salaried Employees	137
D. Gains to Wage-Earners	138
1. Wages	138
2. Wage Differentials	145
3. Hours of Labor	147

CHAPTER VIII
Geographical Shifting of the Industry

The First Location of the Industry	149
The Beginnings of the Industry in Akron	150
Centralization in Akron	152
A. Prior to 1914	152
B. 1914-1929	153
1. The Expansion of the Akron Companies, 1914-1920	154
2. The 1921-1929 Expansion	155
The Beginnings of Decentralization	157
A. The Experience of the United States Rubber Company	157
B. Foreign Branch Factories	159
C. Other Decentralizing Movements, 1920-1929	160
1. Kelly-Springfield	160
2. The West Coast Branches	161
3. The Goodyear Branch in Gadsden, Alabama	162

CONTENTS 11

	PAGE
Trends since 1929	163
A. 1929 to 1935	163
B. The New Movement Toward Decentralization since 1935	166
C. Reasons for Decentralization	171

CHAPTER IX
Summary and Conclusions

Productivity	176
Conditions Attending Increasing Productivity	178
Effects of Increasing Productivity	181
The Outlook for Future Increase in Productivity	187
Other Probable Future Trends	191
A Selective Bibliography	195
Index	201

LIST OF TABLES

		PAGE
I	Physical Volume of Production of the Tire Industry, 1910-1938.	36
II	Average Mileage and Weight of Tire Casings, for Selected Years, 1905-1937	39
III	Tire-Miles Produced in Selected Years, 1910-1937	41
IV	Estimated Production of Automobile Tire Casings by Types, 1910-1933	43
V	Value and Price of Products of the Tire Industry, for Selected Years, 1914-1935	51
VI	Analysis of Principal Markets for Pneumatic Tires, 1910-1936.	54
VII	Practical Physical Capacity of the Tire Industry, 1921-1938 .	60
VIII	Employment in the Tire Industry for Selected Years, 1914-1938	63
IX	Total Labor Productivity per Worker and per Man-Hour in Tire Manufacturing for Selected Years, 1914-1937	69
X	Direct Labor Productivity per Man-Hour in Six Large Tire Plants for Selected Years, 1914-1931	73
XI	Direct Labor Productivity by Departments in Six Large Tire Plants for Selected Years, 1922-1931	76
XII	Value Productivity in the Tire Industry and in All Manufacturing Industry for Selected Years, 1914-1935	77
XIII	Physical Productivity in the National Economy and in the Manufacturing Industries Group in the United States for Selected Years, 1914-1935	80
XIV	Number of Establishments and Number of Wage-Earners per Establishment in the Tire Industry for Selected Years, 1911-1937	88
XV	Labor Costs per Tire and per Tire-Mile for Selected Years, 1914-1937	129
XVI	Wholesale Prices of Tires for Selected Years, 1914-1937	131
XVII	Profits in Tire Manufacturing and in All Manufacturing, 1919-1935	135
XVIII	Wages in the Tire Industry for Selected Years, 1914-1938	140
XIX	Labor Productivity in Tire Manufacturing for Selected Years, 1914-1937	177
XX	Unit Labor Requirements in Tire Manufacturing for Selected Years, 1914-1937	178

CHAPTER I
INTRODUCTION

THE PROBLEM

THE production of the greatest number of desirable goods and services to satisfy human wants with the least expenditure of human effort is a basic problem of our economic society. American manufacturers and business leaders have taken pardonable pride in the fact that remarkably high and rapidly increasing industrial efficiency are outstanding among the characteristics by which American industry is known throughout the world. But the industry of a nation as large as ours is made up of so many complex activities that a detailed analysis of its industrial efficiency represents a very large order indeed. It, therefore, seems probable that more can be learned from a detailed study of specific industries than from a general study of the whole body of industrial operations of the national economy. The present volume is confined to industrial efficiency in the tire industry in the belief that the description of its progress will not only reveal interesting facts with regard to an important segment of the nation's industrial life, but will also contribute something to our understanding of the more general forces which affect the industrial efficiency of our entire economy.

The emphasis in the present investigation has been placed upon recently developed technical measures of the physical volume of production per unit of labor time which are used as criteria of the *productivity of labor*. These methods are widely employed for studying the changes in the effectiveness of the utilization of labor, with regard both to the economy as a whole and to specific industries. Experience thus far seems to indicate that they are most reliable when applied to specific industries which can be clearly defined. Summarily stated the purposes of the present inquiry are: (1) A measurement of the changes in the productivity of labor in the tire manufacturing industry since 1914; (2) An analysis of the conditions attending these changes in productivity and an appraisal of the major factors responsible for them; (3) A description of the principal effects of these changes upon our economic society, including a study of the distribution of the gains resulting from the productivity increase; and (4) An attempt to predict the probable future course of labor productivity in the industry.

Previous Investigations in the Field

The Bureau of Labor Statistics of the United States Department of Labor first developed the techniques for the measurement of labor productivity. The definition of the term and the uses for which it is most appropriate are explained by the Bureau as follows:

"'Productivity of labor' means work done in a given time; ordinarily it is best expressed as output per man-hour, although it may be stated as output per man per day, per crew, per week, etc. The advantage of using man-hour output as a basis of measurement is that it is more precise and exact than the others. The 'productivity' of labor must be clearly differentiated from the 'efficiency' of labor, or any other term which is narrowed down to express only the output due to the ability and willing cooperation of the workers themselves.

"In determining productivity, the laborer is simply used as the unit of measurement in expressing the technical progress or decline of an industry over a period of time, regardless of whether the changes in output were due to new machinery, managerial skill, or better work by the employees.

"The phrase 'technological change' is defined to include all change, whether in the nature of the product, method of production, hours worked, or machinery and equipment used, which results in higher productivity per man-hour.

"Usually the object of technological changes is to reduce the labor cost of operation. This reduction is measured by the difference in the labor requirements per unit of output before and after the change in technology took place, which may or may not result in the immediate elimination of jobs or workers

INTRODUCTION 17

from the plant. It produces a surplus of labor time, and unless there is a corresponding increase in the total output, some workers will eventually be eliminated as a direct result of the technological change. This condition, which the Bureau calls ' labor displacement,' is frequently referred to also as 'technological unemployment'." [1]

The nature of the concept of labor productivity occasionally has led to misinterpretation of the meaning of the measures resulting from productivity studies. The productivity of labor is simply one way of measuring the productivity of industry. The mere fact that production is expressed per unit of labor does not imply that the laborers are responsible for increasing productivity any more than the expression of output per unit of capital invested would imply that capital is responsible for the advance. In both cases the productivity measurements are simply useful statistical methods of stating certain trends in the industry. These trends are the results of the interaction of many complex factors, which are described and analyzed in the latter part of this volume.

The Bureau of Labor Statistics has been making studies of the productivity of labor for a period of at least twenty years during which time more than fifty investigations relating to about forty industries have been made.[2] In one such study the production per person engaged in the rubber industries as a whole was shown to have increased 139 percent between 1909 and 1925, and 99 percent between 1919 and 1925. This study, one of the first to call attention to the fact that the rubber industries were characterized by an exceedingly rapid rate of productivity increase, revealed that the motor vehicle industry was the only one which surpassed the rubber group in the rate of productivity increase between 1919 and

[1] *Handbook of Labor Statistics*, 1936 Edition, United States Bureau of Labor Statistics, Bulletin No. 616, Washington, Government Printing Office, pp. 709-710.

[2] Elizabeth A. Johnson, *A Selected List of the Publications of the Bureau of Labor Statistics*, 1936 Edition, United States Bureau of Labor Statistics, Bulletin No. 624, Washington, Government Printing Office, pp. 19-21.

1925. It was also suggested that the favorable showing of the rubber industries group in this respect was in large part due to the rapidly increasing productivity in tire manufacturing.[3]

In 1926 the tire industry was added to the list of industries studied separately by the Bureau, and man-hour productivity data were computed for the period 1914 to 1925.[4] These indexes were subsequently revised slightly and carried through 1927 by the Bureau,[5] and in a further study by Dr. Arthur F. Beal the data were revised once more and extended through 1931. The data as they appeared at this time were as follows:[6]

Year	Rubber Tires and Inner Tubes Productivity per Man-Hour Index
1914	100
1919	168
1921	221
1923	287
1924	332
1925	336
1926	345
1927	367
1928	404
1929	391
1930	421
1931	512

[3] "Digest of Material on Technological Changes, Productivity of Labor and Labor Displacement," *Monthly Labor Review*, Volume XXXV, November, 1932, pp. 1-27.

[4] "Productivity of Labor in the Rubber Tire and Iron and Steel (Revised) Industries," *Monthly Labor Review*, Volume XXIII, December, 1926, pp. 28-34.

[5] "Productivity of Labor in Eleven Manufacturing Industries," *Monthly Labor Review*, Volume XXX, March, 1930, pp. 501-517.

[6] "Dispersion in Man-Hour Productivity since 1929," *Proceedings of the American Statistical Association*, Volume XXIX, March, 1934, Supplement, pp. 66-71. Although this study is a revision and extension of the study presented in the *Monthly Labor Review* of March, 1930 (*op. cit.*), it is not an official publication of the Bureau but is solely the work of Dr. Beal.

The production data were obtained from the Rubber Manufacturers Association in terms of the number of tires and inner tubes produced. The production index was a weighted average of the various types of products. The weights were based upon the values of the products as shown by the Census of Manufactures. Employment data were secured from the Census of Manufactures and the Bureau of Labor Statistics and were multiplied by average hours per week to secure man-hours. The average hours per week were secured for the most part from the Census of Manufactures and the United States Department of Commerce (*Survey of Current Business*) supplemented by one field investigation of wages and hours in the industry.[7]

These pioneering efforts showed that the industry was characterized by a rapid rate of productivity advance. They were, however, subject to considerable margins of error, as is indicated by the fact that the frequent revisions of the data resulted in substantial changes in the series from time to time. The employment and production data used were not strictly comparable, especially for the years prior to 1921; the man-hour data were based upon average hours per week which were derived by a round-about process; and important changes in the quality of the tire made the production index increasingly unreliable. Recognizing the inadequacies of the data in these forms the Bureau of Labor Statistics in 1932 commissioned Mr. Boris Stern to make a field survey of the industry in order to determine more precisely the developments in productivity in tire manufacturing.[8] This study represents the greatest contribution to our knowledge of the trends in labor productivity in the tire industry to date, and since considerable reliance has been placed upon it in later sections of this volume the results obtained by Stern are only very briefly summarized at this point.

[7] *Wages and Hours of Labor in the Automobile Tire Industry, 1923*, United States Bureau of Labor Statistics, Bulletin No. 358, April, 1924, Washington, Government Printing Office.

[8] "Labor Productivity in the Automobile Tire Industry," United States Bureau of Labor Statistics, Bulletin No. 585, July, 1933, Washington, Government Printing Office.

Stern found that nearly 7 times as many pounds of tires and 5½ times as many tires were being produced per direct labor man-hour in 1931 as in 1914, and that between 1922 and 1931 the productivity per direct labor man-hour increased nearly 2¾ fold in terms of pounds of tires and nearly doubled in terms of number of tires. These data indicate that the rate of productivity increase in the tire industry was even more rapid than had been shown by the previous studies and that, for a period of at least ten years, productivity had been increasing in the tire industry in both good years and bad. Stern also provided an analysis of the productivity increase by departments of tire manufacture, and compiled a large amount of material relating to specific causes and effects of the changes in productivity.[9]

A more recent study of the Bureau of Labor Statistics, in which published sources rather than a field investigation were relied upon, demonstrates that the increase in pounds of tires per man-hour between 1932 and 1936 amounted to 16 percent. This study shows that the rate of productivity increase in the tire industry has been greatly reduced from that of former years and, further, that between 1932 and 1936 the productivity increase in the tire industry was at a rate only one percent greater than in manufacturing industry as a whole. The tire industry, which was among the leaders in productivity increase for more than a decade prior to 1931, has been surpassed by many other industries in this respect since 1932.[10]

Unfortunately, this study, although in many ways the most accurate one on the subject, is not comparable with the previous investigations of the Bureau. It is on a poundage basis whereas the earlier studies published in the *Monthly Labor Review* measured production and productivity in terms of the number of tires. Bowden's study differs from Stern's chiefly because

---

9 See pp. 72-76 for the more detailed treatment of Stern's results. Stern's procedures and the distinction between direct and indirect labor are also discussed at that point.

10 Witt Bowden, "Labor in Depression and Recovery, 1929 to 1937," *Monthly Labor Review*, Volume XLV, November, 1937, pp. 1-37.

INTRODUCTION 21

the latter was confined to direct labor but the former included all wage-earners in the factories.

Because of the lack of comparability of the several previous investigations, the present study undertakes to prepare an independent and comprehensive index of labor productivity in the tire industry.

METHODS AND PROCEDURES

In the preparation of these measurements heavy reliance has been placed upon several of the concepts and methods employed by the Bureau of Labor Statistics. The definitions of the productivity of labor and of technological change as used by the Bureau have already been cited.[11] Virtually complete production and employment data for the industry have been obtained in comparable terms from published sources. In some cases, however, conversions of products to equivalent units have been required. The primary employment and productivity measures relate to the combined direct and indirect manufacturing labor on an annual basis. The physical volume of production has been measured in several ways, including number of tire casings produced, output of the tire industry in terms of equivalent number of tire casings, crude rubber consumed, and tire-miles. The equivalent number of tire casings represents an adjustment to take into consideration the production of other articles in tire plants. This conversion was made on the basis of the value of the various products as given by the Census of Manufactures. The crude rubber consumed and tire-mileage bases take into account the changes in tire quality. The number of persons employed in the industry was directly available for the entire period, and man-hours were derived by multiplying employment by average hours worked per week. The employment data were obtained from the Census of Manufactures and the Bureau of Labor Statistics, and average hours per week were secured from the latter source and the National Industrial Conference Board.

11 Pp. 16-17.

Thus it has been possible to construct six measures of productivity: (1) tires per man-year; (2) tires per man-hour; (3) pounds of rubber consumed per man-year; (4) pounds of rubber consumed per man-hour; (5) tire-miles per man-year; and (6) tire-miles per man-hour.[12]

These productivity measures are in physical terms and are measures of the average productivity of manufacturing labor in the industry. The concepts are dynamic in that the chief value of the measures is to indicate changes in the productivity of labor over time. The measures of productivity used herein are, therefore, to be sharply distinguished from the marginal productivity concepts of economic theory by which the proportions of the factors of production and the returns to individuals and to factors are determined in a given situation at a given time.

The statistical results of the application of the techniques described above provide productivity measures that are sufficiently exact to form the basis for the subsequent discussion and interpretation of productivity movements. Statistics are ways of conveying a complex history in brief and easily understandable form, but the really significant factors are the complex changes lying behind the statistical measurements. Due to the difficulty of arriving at what may be regarded as fundamental causes of productivity changes, the factors involved are discussed as conditions attending increasing productivity.

The approach is descriptive and statistical so far as possible, but a large reliance has been put upon the first-hand observations of the author in his several trips through individual factories and upon interviews with a number of executives, managers, research men, and workmen in the tire industry. In addition use has been made of special studies of specific prob-

[12] The derivations of the statistical series used, the methods employed, and the assumptions and limitations are described in greater detail as the data are presented in later chapters. The statistical data have been supplemented by a number of interviews and have been checked for reasonableness by persons in the industry.

lems in the industry made by individual companies, trade associations, labor unions, government departments, local civic associations, and private statistical agencies. Interviews with persons in these various agencies who qualify as expert observers of the industry have also been found most helpful.

The treatment of the effects of the changes in labor productivity in the industry, which forms the third phase of the investigation, is similar to that followed in the investigation of the conditions attending the changes. Published and unpublished statistical and non-quantitative historical data were relied upon heavily for the financial history and the changes in costs, profits, and wages, but much valuable information was obtained from interviews. The opinions of qualified observers both within and outside of the industry have been canvassed in several instances in which other information was not available. The conclusions arrived at in this manner have been used only when there was substantial agreement among the observers. The analysis of the distribution of the gains resulting from the increases in productivity, which forms a large part of the section on the effects of increasing productivity, is largely but by no means entirely statistical. In connection with the study of the effects of the changes in productivity a separate chapter has been included on the geographical shifts in the location of the industry. The reasons for the shifting of the industry are too complex to permit the phenomenon to be regarded as entirely a result of productivity or labor cost differentials, but these factors certainly loom large in the situation. In any case the subject is one which may be better understood in connection with the developments in productivity. The recent tendency toward decentralization of the industry constitutes a reversal of a trend which has persisted for more than two decades, and although it is an exceedingly timely subject no systematic analysis of the movement has as yet appeared.

In the final chapter the most significant results have been summarized and an effort has been made to predict the probable course of productivity in the industry in the near future.

Although the elements involved are numerous and complex and the disaster which has overtaken many attempts at prediction in the realm of economic developments makes one hesitate, it is believed, none the less, that a few generalizations with regard to future, as well as to past, tendencies have been established.

A study of labor productivity in an industry, especially when combined with an analysis of the attending conditions and effects of the changes, necessarily leads the investigator into a number of side issues. Because of the emphasis on the main theme, however, other questions have been treated only in so far as they are essential to an understanding of the specific phases of the productivity investigation.

CHAPTER II

THE ORIGINS AND ECONOMIC CHARACTERISTICS OF THE TIRE INDUSTRY

Emergence of the Tire Industry

ALTHOUGH crude rubber has been known to the natives of the American tropics for centuries and was introduced to Western Europe more than four hundred years ago, it remained merely a scientific curiosity until Charles Goodyear discovered the process of vulcanization in 1839. The rubber manufacturing industries were thus solidly established and grew steadily in the latter part of the nineteenth century. Among the hundreds of products manufactured, rubber boots and shoes, rainwear, and mechanical rubber goods were of greatest importance.[1]

The search for new uses of rubber has been under way since the substance was first encountered by the white man and particularly since the discovery of the vulcanization process. One of these innovations was the manufacture of solid rubber tires for carriages, which Whittlesey records as a product of the industry as early as 1856.[2] The product apparently did not enjoy a wide distribution for some time, however, because Harvey Firestone states that in 1896 he had the only rubber tired buggy in Detroit.[3] Nevertheless within the next few years, a number of producers, including Firestone, were manufacturing such tires and enjoying an expanding business. It was largely out of the business of carriage tire manufacturing that the tire industry

---

[1] Charles Goodyear, *Gum-Elastic and Its Varieties*, Volumes I and II, New Haven, Conn., printed privately 1853 and 1855. Reprinted by *The India Rubber Journal*, London, England, 1936 and 1937.

[2] Charles R. Whittlesey, " Rubber," *Encyclopaedia of the Social Sciences*, New York, The Macmillan Company, 1934, Volume XIII, pp. 453-461. Geer speaks of their use on carriages in London in 1861. W. C. Geer, *The Reign of Rubber*, New York, The Century Company, 1922, p. 153.

[3] Harvey S. Firestone and Samuel Crowther, *Men and Rubber*, New York, Doubleday, Page and Company, 1926, p. 29.

grew, but this growth came as a result of two other important developments, the invention of the pneumatic tire and the emergence of the automobile as a commercial product.

The principle of the pneumatic tire was first patented by Robert William Thomson in England in 1845. It contained the essential features of the modern tire in that it consisted of a stiff outer casing made of rubber and fabric and an inner tube of softer rubber, which was inflated with air. Thomson's tires were made for carriages and an early set ran 1,200 miles, but they won no general acceptance and the invention was forgotten for nearly half a century.[4] The principle was rediscovered independently by John Boyd Dunlop, a veterinarian of Belfast, Ireland, when he made pneumatic tires for his little boy's tricycle in 1887. The invention was patented in 1888, and the company which was formed to exploit the invention in 1890 has since grown to one of the world's largest tire manufacturers (The Dunlop Rubber Company, Ltd.).[5] The new tires were introduced on bicycles in the 1890's and they were in considerable part responsible for the wide vogue which bicycles enjoyed in that period. Pneumatic tires proved to be much lighter, faster and easier riding than the solid rubber tires, which they soon displaced, and by 1900 several rubber manufacturers were making bicycle and carriage tires a considerable part of their business. Most of the carriage tires were still of the solid rubber type, but bicycle tires were nearly all pneumatic by the turn of the century.

In the meantime the automobile was emerging as a practical commercial product. The historical background of this highly

[4] Wallace H. Paull, "Tyre," *The Encyclopaedia Britannica*, Fourteenth Edition, 1929, Volume XXII, pp. 653-656.

[5] *Ibid.* See also " John Boyd Dunlop," *The Encyclopaedia Britannica*, Fourteenth Edition, 1929, Volume VII, p. 743 and A. T. Fidler, "Dunlop Rubber Company, Ltd.," *The Encyclopaedia Britannica*, Fourteenth Edition, 1929, Volume VII, p. 743. Several patents in addition to Thomson's had been issued prior to Dunlop's and rival claimants in his own time appeared. The Dunlop Company, however, held several accessory patents which enabled it to establish its position.

ECONOMIC CHARACTERISTICS 27

complicated triumph of the machine age includes most of the scientific, technical, industrial, and commercial developments since the steam engine. Various inventors in Europe and America late in the eighteenth and throughout the nineteenth century produced self-propelled vehicles to run on roads. The early ones were steam-driven, but in the latter part of the nineteenth century a number of such vehicles were produced with internal combustion engines using gasoline for fuel. By 1890 the automobile was emerging from the stage of early experimentation and in the first decade of the twentieth century automobiles were being manufactured for sale.[6] The first such cars were almost all produced to order for specific buyers, but in 1908 Henry Ford produced the first Model T and in the following year he announced his famous policy of producing standard low-priced cars on a quantity basis. In 1908 The General Motors Corporation was formed as a merger of several producers and this company soon adopted the same policy. The automobile industry experienced a period of enormous growth as is indicated by the following production and registration figures.[7]

Year	Production	Registrations
1895	4	4
1900	4,000	8,000
1910	187,000	469,000
1920	2,140,000	9,296,000
1930	3,357,000	26,737,000
1936	4,454,000	28,461,000

[6] In November, 1900, the first automobile show was held in New York. It featured no less than 33 automobile and 8 tire exhibitors. Geer, *op. cit.*, p. 137.

[7] *Automobile Facts and Figures*, 1937 Edition, Automobile Manufacturers Association, Inc., New York, pp. 4, 9, and 16. The data relate to the total production and registrations of passenger cars and motor trucks in the United States. Certain adjustments have been made by the present author to eliminate Canadian figures from the production and to add the tax-exempt official cars of the government to the registration figures. The data for these adjustments were obtained from the above publication, pp. 9 and 16. See also Charles Franklin Kettering, "Motor Car", *The Encyclopaedia Britannica*, Fourteenth Edition, 1929, Volume XV, pp. 880-897.

In the nineteenth century several of the early automobiles were equipped with solid tires of the type which were beginning to be used for carriages. It is not clearly established who first made pneumatic tires for automobiles, as several companies claim this distinction, but it is known that many sets were made before the close of the nineteenth century. The first automobile tires were adaptations of bicycle tires, and most of the early ones in America were of the single tube type, consisting merely of a round piece of rubber tubing of nearly even strength and thickness.[8] In England and France the double tube type, consisting of an outer casing and inner tube, was being made in the late 1890's and the early 1900's, and this type soon won general acceptance in America also.[9]

One of the most perplexing problems of the early tire manufacturers concerned the method of fastening the tires securely to the rims. The early solid tires were wired on, and for some time each maker had his own design of both tires and rims. Many ingenious devices for attaching the first pneumatics of both the single and double tube type were experimented with, but by about 1903 the clincher type of tire and rim was standardized by an agreement among a group of manufacturers known as the G. and J. Clincher Tire Association. This group held the basic patents which were supposed to cover all types of clincher tires. They exploited their monopoly by standardizing prices, licensing the patents among themselves, and refusing to admit outsiders. Their reign was ended in 1907, however, when their basic patent was held invalid.[10]

The general acceptance of the double tube clincher type, which had occurred by 1905, was followed by more than two decades of remarkable expansion. In 1906 automobile registra-

---

[8] Firestone and Crowther, *op. cit.*, p. 77. These tires were known as the Tillinghast type and were invented for use on bicycles by Pardon W. Tillinghast in the United States in 1890. *Rubber, Its History and Development*, Firestone Tire and Rubber Company, Akron, 1922, p. 19. Some tires of this type remain in use upon bicycles.

[9] Firestone and Crowther, *op. cit.*, p. 77.

[10] *Ibid.*, pp. 77-89.

tions in the United States first passed the 100,000 mark but they exceeded 1,200,000 in 1913.[11] Automobile tires were merely a small line of rubber manufacturing in 1905. By 1914 the industry had grown tremendously, largely because of the expansion of tire demand. Tires accounted for 49 percent of the value of products of the rubber industries in 1914 and 66 percent in 1919.[12]

The manufacture of automobile tires may be said to have emerged as a separate industry by 1914. Since that time many factories, including most of the larger producers of rubber goods, have been primarily tire producers and their other products have been sidelines. This development was recognized by the Census of Manufactures somewhat belatedly, however, so the data prior to 1921 present some difficulties of interpretation.

The Census collected data for rubber manufactures from 1879 to 1919 under the title " The Rubber Industry " which included several sub-classifications. Until 1914 tires and tubes were included under the " Rubber Goods, Not Elsewhere Specified " group. In that year tires and tubes represented 65 percent of the total value of products of the group.[13] Recognizing the dominance of tires, in 1919 the Census changed the name of this sub-classification to " Rubber Tires, Tubes and Other Rubber Products " without changing its content. Tires and tubes accounted for 76 percent of the value of products of this group in 1919.[14]

In 1921 the Census altered considerably its basis of classification. The establishments formerly grouped under "The Rubber Industry," were divided into three separate industries

[11] *Automobile Facts and Figures*, 1937 Edition, p. 16.

[12] United States Department of Commerce, Bureau of the Census, *Census of Manufactures, 1914, The Rubber Industry*, Washington, Government Printing Office, Table 12; *ibid.*, 1919, Table 11.

[13] *Census of Manufactures, 1914, The Rubber Industry*, Tables 2 and 12. See also Table V of the present volume.

[14] *Census of Manufactures, 1919, The Rubber Industry*, Tables 1 and 11. See also Table V of the present volume.

30   THE TIRE MANUFACTURING INDUSTRY

entitled " The Rubber Industries." The " Rubber Tires and Inner Tubes " industry, which forms the largest of these has been retained in comparable form since 1921. The Census defines it to include " those establishments which are engaged primarily in the manufacture of pneumatic tires, inner tubes, and solid and cushion rubber tires, for any class of vehicles." [15] Since 1921, establishments producing less than $5,000 worth of products per year have been excluded. This limitation on the data is insignificant, however, because in the three preceding Censuses the establishments in this category produced less than one-tenth of one percent of the value of products of the industry.[16] In 1925 only three-tenths of one percent of the tires and inner tubes manufactured in the United States were produced in plants classified by the Census in other industries.[17] The tire industry does, however, manufacture a wide range of other goods, although tires are in all cases the major product of the plant. Since 1921 the value of non-tire products has averaged about 15 percent of the total for the industry.[18]

THE ECONOMIC CHARACTERISTICS OF THE TIRE INDUSTRY

A. YOUTH

The most outstanding characteristic of the industry is still its relative newness. From 1910 to 1920 production expanded at a truly enormous rate showing a somewhat more than 20-fold increase. During the next decade it continued to expand but the rate of increase was slowing down. Production was sharply curtailed between 1929 and 1932, and although it has been increasing since the latter date, the industry now appears to have entered a period of relative maturity characterized by only a moderate rate of growth. The slower rate of growth is

---

15 *Census of Manufactures, 1929, The Rubber Industries*, p. 1. This is the definition of the industry used in the present study.

16 *Census of Manufactures, 1919, The Rubber Industry*, Table 7.

17 *Census of Manufactures, 1925, The Rubber Industries*, Table 4. Practically all of these were in the other rubber industries.

18 See Table V.

ECONOMIC CHARACTERISTICS 31

largely the result of two factors: (1) the slackening in the rate of growth of the automobile industry, and (2) the remarkable improvement in the wearing quality of tires. The relationship between the growth cycle of the industry and the changes in the productivity of labor will be elaborated in a later section.

B. TIRE DEMAND

Tires are usually regarded as consumers goods inasmuch as tires for passenger cars represent the major share of the industry's output.[19] Until recently tires were also generally considered to be non-durable, but the improvement in quality has been so great that they must now be considered at least semi-durable. Automobile ownership has become so widespread in the United States in recent years that cars and therefore tires have dropped out of the luxury class of commodities, and have come to be regarded as necessary conveniences.[20] The domestic market absorbs from 95 to 98 percent of the total tires sold. The most important marketing characteristic of tires, however, is that they are strictly a product of joint de-

---

[19] In 1935 42,479,133 passenger car tires were produced at a wholesale value of $221,555,480. In the same year the industry produced 6,003,338 pneumatic truck and bus tires with wholesale value of $99,852,136. The industry also produced 435,884 tires for airplanes, industrial trucks, tractors, and other industrial uses with a total wholesale value of $3,173,396. Thus, although about 87 percent of the tires produced were for passenger cars, they represent only a little more than two-thirds of the value of the industry's production. *Census of Manufactures, 1935, The Rubber Industries*, Table 4.

[20] A United States Department of Commerce survey of car ownership by income classes in 18 cities in 1933, showed that 54.2 percent of all families owned cars. The proportion ranged from 29 percent for families having incomes from $0 to $499 to 92 percent for families having incomes of $5,000 or more. Only 2.8 percent of the families owned two or more cars and one quarter of one percent of the families owned three or more cars. *Automobile Facts and Figures*, 1937 Edition, p. 85.

A larger proportion of the families in small towns and rural areas own cars than in the cities so it may be estimated that at least two-thirds of the American families own cars. In 1936, there was a passenger car for every 5.3 persons in the United States and a motor vehicle for every 4.6 persons, ranging from one for every 2.6 persons in Nevada to one for every 9.8 persons in Mississippi. *Ibid.*, p. 21.

mand. Tires are, moreover, only a relatively small part of the cost of automobile transportation. The average motorist spends approximately $180 annually for this service and of this total only about $17.50 is spent for tires.[21] This circumstance more than any other has tended to make automobile tires a product of relatively inelastic demand. Prior to 1929 this factor, together with the rapid growth of demand, was considered to make the tire industry relatively unsusceptible to general business fluctuations. Improvements in wearing qualities have made tires a more postponable type of purchase, however, so that the above generalization is now less applicable than it was a decade or more ago.[22]

C. TIRE SUPPLY

The relative newness of the industry and its growth cycle are, of course, primarily characteristics affecting supply. Another feature is the absence of competing industries supplying the same essential demand, although the automobile and related industries compete with other forms of transportation. The tire industry is commonly regarded as a mass production industry, that is, it produces large quantities of standarized articles using large amounts of capital, highly mechanized processes, and minute divisions of labor into repetitive tasks. The industry has frequently been compared with the automobile industry in these respects but it differs from that industry in being much smaller and less mechanized and in requiring a greater proportion of skilled workers. However, in spite of the fact that there

---

[21] W. W. Leigh, "The Wheels of Fortune," *Tire Review*, February, 1938, pp. 10-18. The $180 average annual expenditure relates solely to the cost of running a car since it excludes depreciation, repairs, garaging, and insurance.

[22] Walton Hamilton and Associates, *Price and Price Policies*, New York, McGraw-Hill Book Company, 1938, Section III "The Automobile Tire — Forms of Marketing in Combat" by Albert Abrahamson, pp. 83-116. See also Leonard E. Carlsmith, *The Economic Characteristics of Rubber Tire Production*, New York, Criterion Linotyping and Printing Company, Inc., 1934.

The development of retreading has also helped to make tire purchases more postponable. See pp. 49-50 of the present study.

are a score of industries having a larger total number of wage-earners and value of sales than the tire industry, relatively few manufacturing plants in the nation exceed the size and employment of the largest tire factories. In 1933, the four largest companies had slightly more than 64 percent of the total capacity of the industry, the next four companies in size had 18 percent of the industry's capacity, and the remaining 18 percent was scattered among some 20 odd other producers.[23] In 1929 two-thirds of the wage-earners in the industry were employed in the 7 plants employing 2,500 or more workers, and 82 percent of the total employment was in the 16 plants employing 1,000 or more wage-earners. Although more than half of the plants employed 250 workers or less they accounted for only 4 percent of the total employment in the industry.[24]

In spite of the high degree of concentration of the industry it has long been one of the most competitive industries in the country. Price wars have been frequent and the trend of prices has been rapidly downward since the beginning of the industry. For a number of years prior to 1920-1921, it was an exceedingly profitable industry but since that time its earnings have been below the average for American manufacturing industries. Nevertheless it is and has been for many years one of the highest wage industries in the country. It is also characterized by a relatively large capital investment per worker and per dollar of sales and, as will be seen from data presented in Chapter V, it has a relatively high value of product and value added by manufacture per unit of labor. The industry has been non-union throughout the greater part of its history, but since 1933 unionization has developed and since 1936 a large proportion of the

[23] Brief submitted to the National Recovery Administration by the Rubber Manufacturers Association (Unpublished). Table shown in W. H. Cross, *The Rubber Tire Manufacturing Industry*, National Recovery Administration, Division of Review, Evidence Study No. 36, Washington, October, 1935, p. 5 (Mimeographed).

[24] A. L. Kress and C. A. Pearce, *Material Bearing on the Rubber Tire Industry*, National Recovery Administration, Division of Research and Planning, Washington, November 9, 1933, p. 34 (Mimeographed).

employees of the industry have been organized in the United Rubber Workers of America, an affiliate of the Congress of Industrial Organizations. The industry as it grew up largely localized in Akron, Ohio, which became known as the rubber and especially the tire center of the United States, but within recent years Akron has been losing its domination of the industry. Akron is still much the largest tire manufacturing center of the nation, however, and produces more than 35 percent of the industry's total number of tires and a somewhat larger proportion of the total value of its products.[25]

[25] The foregoing paragraph is intended to give the reader only a very general description of some of the outstanding characteristics of the industry. Supporting data, together with a greater amount of detail, will be found in subsequent chapters.

CHAPTER III

PRODUCTION

Tires Produced and Rubber Consumed

The tire industry is considered to have emerged as a separate industry by 1914 and, therefore, the greater part of the discussion in this and the following chapters concerns its development since that time. Since production data in physical terms are of the greatest significance from the point of view of productivity, they have been given primary emphasis. A large share of the history of the industry is to be found in these production figures. The physical volume of the output of the industry has been measured in three ways, the details of the movements of which are brought out in Table I. The first and most obvious of these measures is the number of pneumatic casings produced in each year. Because of the increasing importance of tires relative to the industry as a whole prior to 1921, however, this measure somewhat overstates the growth of the industry in that period. In order to obviate that difficulty an index of the total volume of production of the tire industry as a whole in terms of equivalent number of tire casings has been constructed. For this purpose tire casings and other products of the industry have been weighted according to their relative values, as shown by the Census of Manufactures.[1] On this basis the physical output of the industry increased nearly five-fold between 1914 and 1925 and showed a further increase of 24 percent over the latter year in 1928. The drop between 1928 and 1932 was unusually severe, production having been cut nearly in half. By 1936 production was approaching the 1925 level, although still 22 percent below the 1928 peak, but in 1937 it fell again to seven percent below 1925 and 25 percent below 1928. In 1938 fewer tires were produced than in any year since 1932.

[1] The weight factors used are shown in Table V.

TABLE I
PHYSICAL VOLUME OF PRODUCTION OF THE TIRE INDUSTRY, 1910-1938

Year	Pneumatic Tire Casings [a] (Thousands) (1)	Indexes: 1925 = 100		
		Pneumatic Tire Casings (2)	Total Production [b] (3)	Rubber Consumed [c] (4)
1910	2,400	4	—	—
1911	3,000	5	—	—
1912	5,000	9	—	—
1913	6,000	10	—	—
1914	8,021	14	21	—
1915	12,000	20	—	—
1916	18,565	32	—	—
1917	25,836	44	—	—
1918	23,000	39	—	—
1919	32,836	56	65	—
1920	32,400	55	—	—
1921	27,298	46	43	39
1922	40,930	70	67	—
1923	45,426	77	77	66
1924	50,820	87	86	82
1925	58,784	100	100	100
1926	60,120	102	101	92
1927	63,550	108	105	91
1928	75,527	129	124	110
1929	69,765	119	114	115
1930	51,610	88	83	90
1931	49,143	84	77	83
1932	40,286	68	64	74
1933	45,376	77	74	89
1934	47,233	80	78	98
1935	48,765	83	82	106
1936	58,091	99	97	126
1937	55,402	94	93	131
1938	40,720	69	—	—

[a] Data for 1910-1935 from E. G. Holt, *United States Renewal Tire Market Analysis*, United States Bureau of Foreign and Domestic Commerce, Rubber Section, Circular Ru-3500, Revised Copy, December 30, 1937, Washington, (Mimeographed) p. 14, Table III; 1936 to 1937 data from the *Survey of Current Business*, United States Bureau of Foreign and Domestic Commerce. The data were originally secured by the Bureau from the Rubber Manufacturers Association.

[b] Index of total production of the tire industry in terms of equivalent number of tire casings. Derived by weighting other products and casings

PRODUCTION   37

IMPROVEMENT IN QUALITY

Such a measure of production, in terms of the number of tires produced annually, is useful in the interpretation of some trends within the industry, chiefly because it is the customary measure of production in physical terms. It does not of itself afford an accurate measure of the physical production of the industry, however, because the typical tire has undergone fundamental changes in durability, weight and size over the period covered by this study. A more satisfactory measure of physical production is, therefore, to be found in Column 4 of Table I, where production is measured in terms of the annual consumption of crude rubber in the manufacture of tires and tubes. Although this measure is conservative, since it considers only one of the major factors involved in the tremendous improvement of the tire quality, it nevertheless shows a very considerable divergence in movement from the index of equivalent number of tires. From 1921, the earliest date for which such data are available, to 1929 the rubber consumed index shows only a slightly greater increase than that based upon the number of tires, but after the latter date the divergence

according to their relative values as shown by the Census of Manufactures. (See Table V for weighting factors.)

An illustration of the method of weighting may serve to clarify the procedure involved. In 1925 58,784,000 pneumatic tire casings were produced. Referring to Table V it is seen that pneumatic tire casings accounted for 71 percent of the value of products of the industry in that year and inner tubes, solid and other tires, and other products represented 29 percent. Dividing 58,784,000 by .71 gives 82,794,000 as the equivalent number of casings. That is to say, if 58,784,000 tire casings represent 71 percent of the value of products of the industry, the output would have been 82,794,000 casings if the industry had produced casings only. Similarly in 1927, 63,550,000 casings were produced, representing 73 percent of the value of products of the industry, so that the equivalent number of casings for that year was 87,550,000.

The proportion of casings to total value of products has been interpolated for intercensus years since 1921. Thus since casings were 71 percent of the total value of products in 1925 and 73 percent in 1927, the figure for 1926 is taken as 72 percent. The 1935 figure (72 percent) has been used for 1936 and 1937.

c Index of the monthly averages of crude (i.e. new) rubber consumed in tires and tubes. Data were taken from the *Survey of Current Business*.

increases rapidly.[2] The depression drop was smaller than indicated by the number of tire casings produced, for by 1935 the index had returned to the level of the mid-twenties, and 1936 and 1937 represented successive all time peaks. Whereas the equivalent number of casings produced in 1937 was seven percent below 1925 and 25 percent below the peak, 1937 rubber consumption was 31 percent above 1925 and 14 percent above the peak of the nineteen twenties. The physical volume of production in terms of rubber consumed has nearly kept pace with the growth of automobile registrations, although the equivalent number of tires produced annually has lagged far behind. These two measures of physical production will be used concurrently throughout the remainder of the study since, although rubber consumed is a more accurate measure, the index based upon the functional unit of the tire is too significant to be entirely disregarded.

The record of the tire industry for continuous and rapid improvement in the quality of its products is even more outstanding than the expansion in the volume of production in terms of either the number of tires or poundage. Within the last thirty years the weight of the average tire has more than doubled, the average life in years almost quadrupled, the average mileage expectancy has increased nearly ten-fold, and it is probable that the ton-mileage expectancy has increased to an even greater extent.[3] These results have been achieved by a

[2] The decline in the rubber consumption index in 1926 and 1927 reflects the efforts of tire manufacturers to economize on crude rubber chiefly by the use of a larger proportion of reclaimed rubber, during the period of high crude rubber prices under the Stevenson Restriction Act. Complete statistics of reclaimed rubber used for tires and tubes are not available, but the indications are that with this exception the proportion did not change materially enough to affect the index. See Howard Wolf and Ralph Wolf, *Rubber: A Story of Glory and Greed*, New York, Covici-Friede, 1936, pp. 340-341.

[3] The increase in mileage is, of course, in part due to the tremendous improvement in roads during the period. The fact that no allowance has been made for the increasing weight of motor vehicles tends to reduce the margin of error somewhat. Holt, moreover, suggests that much of the increase in tire life attributable to better roads has been counteracted by the attendant higher

TABLE II

AVERAGE MILEAGE AND WEIGHT OF TIRE CASINGS FOR SELECTED YEARS, 1905-1937

Year	Average Mileage [a] (1)	Average Weight [b] (lbs.) (2)	Average Life [c] (Years) (3)
1905	2,000	—	—
1910	3,000	—	0.73
1915	3,500	12.8 [d]	0.77
1920	5,000	15.3	1.28
1925	10,000	17.3	1.58
1930	15,000	22.9	2.47
1937	20,000	26.1 [e]	2.69 [f]

[a] Scudder, Stevens and Clark, Investment Counsel, *Report on The Automobile Tire Industry*, Boston, 1931, p. 31 (Multigraphed); Cross, *op. cit.*, p. 14; Stern, *op. cit.*, p. 2. The figure given for 1937 (20,000 miles) has been used by the Rubber Manufacturers Association since the N.R.A. days, and although it was probably a trifle high at that time it may be taken as the average performance for the 1937 tires. These data relate to tires *produced* in the years specified not to tires in use at the time.

[b] Carlsmith, *op. cit.*, p. 96.

[c] E. G. Holt, *Rubber News Letter*, June 15, 1937, United States Bureau of Foreign and Domestic Commerce, Circular No. 3644, Washington, p. 9 (Mimeographed).

[d] This figure is for 1916. It is the earliest figure given by Carlsmith.

[e] This estimate was made by the present author. It is a compromise between the N.R.A. and the Rubber Manufacturers Association estimates. The last figure given by Carlsmith is 24.2 lbs. for 1933.

[f] This figure is for 1936. 1937 data were not available although a conservative estimate would be slightly in excess of 2.7 years.

series of changes in the size, physical construction, and chemical composition of the tire which have revolutionized the industry.

As shown in Column 1 of Table II, the average mileage of tires increased from 2,000 in 1905, to 10,000 in 1925 and to 20,000 in 1937. The average weight of a tire meanwhile increased very slowly until about 1915, but from that year until the present it has increased at a cumulative rate of about four

average speeds of operation and the more frequent braking required by the increasingly congested condition of the highways (United States Bureau of Foreign and Domestic Commerce, Circular Ru-3500, p. 4).

percent per annum.[4] It appears that the weight of a tire is an important factor in its quality but that it has been by no means the sole factor in increasing tire durability. The average life of a tire in years has increased from about nine months to about thirty-two months since 1910. This factor is of limited applicability in the measurement of tire durability, however, because of the obvious fact that the mileage the average car is driven per year has increased considerably over this period.

Tire-Miles Produced

From the production of tire casings as shown in Table I, and the average mileage of tires as shown in Table II, it is possible to estimate the physical volume of production in terms of tire-miles produced in certain years. These data are presented in Table III.

This measure of production does not take into account the products of the tire industry other than pneumatic tire casings, the improvement in the quality of which was on the whole less rapid during the period than in tire casings themselves. Moreover, it assumes that the entire improvement in the durability of tires was a result of the tire manufacturing process, whereas we have seen that this result was probably in part due to better roads. As a consequence the increase in production is somewhat overstated by the tire-mileage basis.

Measuring production in terms of tire-miles indicates a much more rapid increase in the physical output of the industry than the measures presented in Table I, and is perhaps a better indicator of the growth of the industry. In the twenty-seven year

[4] Carlsmith, *op. cit.*, p. 96. Dr. Carlsmith arrived at this estimate by fitting a trend to the average weights for the period 1916 to 1933. There is some indication of a slowing down of the rate of increase since the latter date.

It is interesting to observe in this connection that, although inner tubes have increased in durability at a rate which if anything has been somewhat more rapid than that of tire casings, there has been no such corresponding increase in their weight. The average weight of inner tubes has remained at between two and one-half and three pounds throughout the greater part of the period, although there appears to have been a tendency toward heavier tubes since about 1929. See Carlsmith, *op. cit.*, pp. 96-97.

PRODUCTION

TABLE III
Tire-Miles Produced in Selected Years, 1910–1937

Year	Estimated Tire-Miles [a] (Billions)	Index Tire-Miles (1925 = 100)
1910	7	1.2
1914	28	4.8
1919	154	26.2
1921	150	25.5
1925	588	100.0
1929	1,046	178.0
1932	705	119.9
1935	921	156.7
1937	1,108	188.5

[a] Production of tire casings (Table I, Column 1) multiplied by average miles (Table II, Column 1). Average mileage data for the years not appearing in Table II were: 1914, 3,500; 1919, 4,700; 1921, 5,500; 1929, 15,000; 1932, 17,500; and 1935, 19,000.

interval between 1910 and 1937, the number of tire-miles produced annually increased no less than 157-fold. The tire-mileage output of the industry has nearly doubled since 1925, and the tire-mileage output of the industry in 1937 was in excess of that of 1929 by some 62,000,000,000 tire-miles, although the output of 1937 was achieved by the production of some 14,000,000 fewer tires.

Factors Responsible for the Increase in Tire Durability

A. GENERAL FORCES

One of the principal expressions of competition within the industry has been the effort of each manufacturer to put out a superior product. This rivalry has been one of the chief forces responsible for the remarkable improvement in tire durability. The newness of the industry, the expansion of the market, and the relatively unstandardized nature of the products have also been powerful stimulants to competition. The unfavorable business conditions in the period following 1929, which were accompanied by cheap raw materials, lower tire prices, and

42   THE TIRE MANUFACTURING INDUSTRY

excess productive capacity, increased the competitive struggle. For some time the research department of one leading tire company has borne the slogan " a tire to last as long as a car." That this is not beyond the realm of possibility is indicated by the calculation that, whereas in 1910 the average car could be expected to wear out 7.6 sets of tires in its lifetime, by 1920 4.7 sets would serve, and in 1936 3.6 sets were all that would be required.[5]

B. CHANGES IN TYPE OF TIRES PRODUCED

The improvement in the quality of tires is largely explainable in terms of a series of changes in type and construction, the progress of which is shown in Table IV. The first of these major changes was the shift from the clincher rim to the straight side type.[6] The latter was first introduced about 1905 and won headway slowly at first, constituting only 2 percent of the total in 1910 and 49 percent in 1925. Within the next few years it rapidly won universal acceptance, however, and as

---

[5] These figures were derived by the writer from the average life of automobiles in years (Holt, Bureau of Foreign and Domestic Commerce, Circular Ru-3500, p. 18), and the average life of tires in years (Holt, Bureau of Foreign and Domestic Commerce, Circular 3644, p. 9) as shown below:

Year	Automobile Life	Tire Life	Sets of Tires per Automobile
1910	5.54	0.73	7.59
1920	5.97	1.28	4.65
1936	9.55	2.69	3.55

Automobiles, as well as tires, have been increasing in average life, although at a slower rate. This explains much of the retardation of the trend toward fewer tires per car.

A more accurate result would be obtained by the use of mileage data for both tires and automobiles but such data are not available for automobiles.

[6] Clincher tires were held to the rims by flanges in the tires which fit securely into special slots in the rims. Although tires and rims of this type have not been used on automobiles for many years, they are still used on some bicycles. Straight side tires are held on by means of their inextensible wire enforced edges or beads. In the course of the industry's history there have been many variations of both types. The straight side drop center rim (in which the center is lower than the sides) is now standard equipment and practically all modern tires are made to fit them.

nearly all manufacturers equipped new cars with straight side tires after 1927, the sales of clinchers after that date were principally renewals for Ford Model T's. The shift in rim type practically eliminated rim cutting, once a serious menace to the life of a tire.

TABLE IV

ESTIMATED PRODUCTION OF AUTOMOBILE TIRE CASINGS BY TYPES [a]
1910–1933

(Percentages)

Year	Construction			Rims	
	Fabric	High Pressure Cord	Balloon Cord	Clincher	Straight Side
1910	100	0	0	98.0	2.0
1911	100	0	0	96.7	3.3
1912	99	1	0	93.0	6.0
1913	98	2	0	91.0	9.0
1914	97	3	0	90.0	10.0
1915	95	5	0	89.0	11.0
1916	92	8	0	87.5	12.5
1917	90	10	0	85.0	15.0
1918	85	15	0	80.0	20.0
1919	75	25	0	75.0	25.0
1920	65	35	0	70.0	30.0
1921	60	40	0	65.0	35.0
1922	51.4	48.6	0	58.8	41.2
1923	42.6	55.4	2.0	61.0	39.0
1924	29.7	58.8	11.5	57.2	42.8
1925	14.1	51.8	34.1	50.8	49.2
1926	5.3	47.2	47.5	40.7	59.3
1927	1.5	44.6	53.9	28.6	71.4
1928	0.6	33.0	66.4	19.6	80.4
1929	0	25.2	74.8	12.7	87.3
1930	0	16.9	83.1	6.3	93.7
1931	0	14.2	85.8	2.0	98.0
1932	0	12.2	87.8	0.5	99.5
1933 [b]	0	10.8	89.2	0	100.0

[a] Data from Holt, Bureau of Foreign and Domestic Commerce, Circular Ru-3500, Table V, p. 16.
[b] First nine months only.

The early pneumatic tires were made with a square woven fabric base in which the lengthwise threads were of the same size and the same number per inch as those running crosswise. Not only were the threads continually sawing each other in two, but they generated high temperatures which materially reduced tire wear and increased the danger of blowouts. Consequently the average life of the best fabric tires was only about 4,000 miles and it was an exceptional occurrence to find one good for 5,000 miles. The cord tire, first introduced about 1912, contained only a few light cross threads, just sufficient to hold the much heavier lengthwise threads (cords) together until they were rubberized.[7] Moreover, each cord was surrounded by rubber, making it impossible for adjacent cords to chafe, thereby greatly reducing the heat generated by the tire. For this reason as well as because of their greater resiliency, cord tires almost from the first gave considerably better mileage than the fabric types. They were soon averaging 8,000 miles and a series of improvements gradually increased the mileage to about 12,000. Cord tires were somewhat more expensive than the older type, however, and were slow in winning dominance. It was not until 1923 that the production of cord tires exceeded that of fabric tires, but the movement was accelerated after that date and the latter type had dropped to less than one percent of the total by 1928.

The year in which cord tires first assumed a dominating position was marked by the appearance of the enlarged low pressure type, known as balloon tires. Whereas all previous tires had been of the high pressure type carrying from 70 to 90 pounds of air pressure per square inch for the average sizes, the balloon tires carried only about 30 pounds or less. The new type was much larger and heavier, and it greatly increased riding comfort by the absorption of road shocks. Many troubles were experienced with the early balloon tires and it was thought at first that their chief advantage lay in superior riding comfort,

[7] At the present time many tires are manufactured from cords containing no cross threads at all.

but within two or three years after their adoption they were improved so as to give 14,000 to 15,000 miles. They won a more rapid general adoption than the earlier improved types, so that by 1927 more than half of the production was of the balloon type and by 1937 their use was well nigh universal on passenger cars. By this time they had been developed to give 20,000 or more miles of wear.

Beginning about 1925 or 1926 there was a movement toward the manufacture of a considerable volume of six- and eight-ply tires instead of the usual four-ply tires. The extra plies or layers of cotton cord, gave these tires stronger side walls and greater durability, as they could be used until the tread was practically all worn off without danger of blowouts. The mileage of these tires ranged up to 25,000 or more. This movement apparently reached its peak in 1930, when 26 percent of all tires for passenger cars were of this construction. By 1936 only 11 percent of passenger car tires produced were of this type. Recently there has been some development toward the so-called super-balloon tire, with even lower pressures and smaller rims.[8]

Since many of these changes in tire construction have overlapped each other for long periods, with the improved type only gradually winning dominance, it is clear that the estimates of average tire life given in Table II are somewhat of the hybrid variety, being based in part upon the percentages of the various types produced in the given years. Some variation in average tire-mileage may also be due to shifts in the proportions of the various quality lines of tires, but the lack of adequate data prevents us from taking this factor into account quantitatively. It is known, however, that in the middle of the last decade when the Stevenson Plan had forced the price of crude rubber to unusually high levels, there was some movement among manu-

[8] Another new development is the "blowout proof" inner tube. The Goodyear Tire and Rubber Company has taken the lead in this field with its "Lifeguard" tube, which consists of a second tube within the ordinary tube. Although it is designed to give increased wear, its chief advantage is the added factor of safety from blowouts. The other major companies have also developed "blowout proof" tires of various types.

facturers to cheapen the product by the use of a greater amount of reclaimed rubber and by increasing the proportion of second, third and, in some cases, even fourth quality lines. The former tendency was reversed with the collapse of rubber prices but the latter has continued with only slight signs of abatement. It has served, however, only to retard slightly the continuing improvement in tire quality and it is a byword in the industry that the second line tires of today are superior to the first quality product of five years ago. Moreover, beginning about 1928 there has been a distinct counter tendency toward four-ply tires above standard quality.

These frequent changes in tire sizes and types, in spite of the remarkable improvement in quality occasioned by them, have not been unmixed blessings to the industry or to the consumer. One undesirable result has been an undue diversification of sizes and types of products, which has operated to limit the improvements in industrial efficiency. Although it might appear to the average observer that a dozen or so types and sizes might be sufficient to take into account differences in weight, power and speed of passenger cars, a retail price list of one prominent manufacturer shows some 90 types and sizes for passenger cars and over 200 when different price lines are considered.[9] The number is increased several fold when tires for trucks and buses are taken into consideration. The diversity is in large measure due to the fact that the older types have to be made to supply the renewal demand for some years after they have been discontinued on the current model cars. Likewise many minor variations are made by the automobile manufacturers in their specifications for new equipment. Within the last few years, however, there has been a trend toward standardization and decreasing the diversity of sizes especially for passenger cars. Nearly one-third of all passenger tires sold in 1937 were

[9] W. H. Cross, G. S. Earseman and J. H. Lenaerts, *The Rubber Industry Study*, National Recovery Administration, Division of Review, Work Materials No. 41, Washington, 1936, p. 103 (Mimeographed).

of one size (6.00x16) and a dozen other sizes accounted for more than another third of all such sales.[10]

C. DEVELOPMENTS IN RUBBER CHEMISTRY

Of the many new processes and improved methods which have been developed since the beginnings of the rubber manufacturing industry, only a few of the most outstanding can be mentioned here. The rubber industries were made possible by chemical research which led to the discovery of vulcanization and the development of the art of compounding rubber with various other ingredients. Most of the early manufacturers of rubber goods were also chemists and inventors who were constantly in search of new products and processes, but after the industry obtained its start a fund of pragmatic knowledge and rules of thumb were evolved and largely standardized throughout the industry. Chemists and technical men were employed in routine testing of materials but little use was made of organized scientific research. The practical leaders of the industry were skeptical of the value of such work. Beginning late in the nineteenth century, however, a number of discoveries and improvements in processes were made by these men which proved to be of such great value to the industry that research departments were set up in all of the large factories and in many of the small ones. Since that time this work has been an essential part of all of the major rubber and tire factories, and the achievements of the scientists have been in large part responsible for the remarkable progress of the industry in improving its products and processes and developing new products.

The discovery of the alkali process for reclaiming rubber by Arthur H. Marks of the Diamond Rubber Company in 1899 was one of the first outstanding achievements of organized research and did a great deal toward furthering its general adoption. By this process reclaimed rubber can be obtained at a fraction of the cost of crude rubber.[11] Only a small proportion

---

10 " Relative Popularity of Tire Sizes," *Tire Review*, April, 1938, p. 49.
11 Wolf and Wolf, *op. cit.*, pp. 336-341. These authors state that reclaimed

of reclaimed rubber is used in the best tires, but it plays an important part in the process and results in the manufacture of better as well as cheaper tires. A larger proportion of reclaimed rubber is used in the cheaper grades of tires. For some products, however, reclaimed rubber is as good or better than crude. In the rubber industries as a whole from 20 to 50 percent as much reclaimed as crude rubber is used. Many of the large tire and rubber factories maintain reclaiming plants, but it is also an industry in its own right, carried on by many independent producers.

Vulcanization in itself is a slow process and from Goodyear's day the industry has been experimenting with accelerators to speed it up. Inorganic accelerators of various types were used by Goodyear in his early experiments and they continue to be important today. Organic accelerators, which were first developed by George Oenslager, an assistant to Marks at the Diamond factory in 1906, represent another large advance. They not only greatly speeded up the process but they also improved the quality of the rubber immensely. Some of these compounds were used to increase the tensile strength of the cheaper grades of rubber so as to make them nearly comparable with fine para, the best and most expensive grade on the market. The discoveries of Oenslager paved the way for further research, the results of which are so notable that there are now some 1,500 different materials used in rubber compounding, different ones being used to achieve different properties of the finished product.[12] About 1912 carbon black was introduced as a reinforcing agent, a development which so increased the resistance of rubber to abrasion as to be in large measure responsible for the increased wearing quality of the present-day tire tread.[13] The development of anti-oxidants has done a great

rubber can now be produced at a cost as low as three cents per pound. Crude rubber is now selling at eighteen cents per pound (April, 1940).

[12] Wolf and Wolf, *op. cit.*, pp. 345-356. Carrol C. Davis and John T. Blake (Editors), *Chemistry and Technology of Rubber*, New York, Reinhold Publishing Corporation, 1937, pp. 288-291.

[13] Wolf and Wolf, *op. cit.*, p. 355.

deal to further reduce the tendency of rubber to deteriorate as a result of age or exposure to the weather. Whereas some years ago many tires became unfit for use because of deterioration before being worn out, this seldom occurs today in spite of the fact that present day tires remain in use three or four times as many months as did those of twenty-five years ago.

D. OTHER FACTORS

Improved raw materials have also been in part responsible for the betterment of tires in recent years. The average quality of crude rubber used improved steadily with the shift in the principal sources of supply from wild rubber to plantation rubber prior to about 1922.[14] The Stevenson Plan, especially during the period of very high prices about 1925, encouraged adulteration among native producers for a time, but this result was only temporary.[15] The crude rubber of the present time is considered to be on the average better than at any time in the past. Originally only the long staple sea island cotton was considered suitable for tire fabric but its increasing scarcity led to the use of other varieties, of which Egyptian for a time predominated. The introduction of the cord tire made the shorter staple varieties acceptable, however, and the best grades from the Mississippi valley are now the main source of supply. The cotton used for tires still commands a higher price than the standard grades but the differential has been substantially reduced in recent years.[16] Most of the recent developments in raw materials have contributed more toward decreasing their costs than toward improving their quality.

The growth of the practice of retreading tires has substantially increased the mileage of the tires subjected to this treatment in recent years. This business is largely in the hands of tire dealers rather than manufacturers, but it has vitally affected

---

14 C. E. Frazier and G. F. Doriot, *Analyzing Our Industries*, New York, McGraw-Hill Book Company, 1932, p. 84.

15 Charles R. Whittlesey, *Governmental Control of Crude Rubber*, Princeton, N. J., Princeton University Press, 1931, pp. 104-105.

16 Scudder, Stevens and Clark, *op. cit.*, p. 12.

the industry and probably will prove to be even more important in the future.[17] The process consists of applying more rubber and a new tread to old tires which have worn smooth but still have sound carcasses. The process originated in the 1920's but made slow progress until 1929 in which year it was estimated to have been applied to about 250,000 tires.[18] It was first developed on a large scale in California but soon spread over the nation, and in 1935 it was estimated that about 2,000,000 tires were retreaded.[19] Since that time the business has expanded to the point where leaders in the industry estimate that nearly 4,000,000 tires were retreaded in 1937.[20] The standard guarantee on retreaded tires is now 15,000 miles or 10 months' service.[21] The cost to the consumer per tire-mile is substantially reduced by the use of retreads.[22]

Sales and Markets

The value of tire products increased rapidly in the prewar decade and continued upward to a peak of more than $900,000,000 in 1926, but by 1936 it had declined to only $406,000,000, or less than half the level of ten years earlier.[23] Total dollar sales of the tire industry have not only failed to advance as rapidly as production, but they have declined dras-

---

[17] The increased tire mileage resulting from retreading is not included in the average mileage per tire data presented in Table II. Had this factor been taken into consideration, the increase in average mileage would have been considerably greater since 1930.

[18] Wolf and Wolf, *op. cit.*, pp. 342-344.

[19] *Ibid.*

[20] "Biggest Year Faces Makers and Users of Tire Renewing Equipment," *Tire Review*, January, 1938, p. 1.

[21] "Safe Tire Sales," *Tire Review*, January, 1938, p. 40. Of course, by no means all of the worn tires are suitable for retreading. Dealers now reject about 20 percent of those offered because of damage, wear or age and the poorest ones are not even offered for such treatment.

[22] The safety factor is, however, lowered somewhat, retreaded tires being more likely to blow out in hot weather or at high speeds than are new tires.

[23] *Automobile Facts and Figures*, 1929 and 1937 Editions, p. 15. All sales and price data relate to wholesale values at the factory.

PRODUCTION 51

TABLE V
VALUE AND PRICE OF PRODUCTS OF THE TIRE INDUSTRY FOR SELECTED YEARS, 1914-1935 [a]

	1935	1933	1931	1929	1927	1925	1923	1921	1919 [c]	1914 [c]
Total Value of Products [b]	446,092	299,313	406,283	770,177	869,688	925,002	644,194	496,123	987,088	223,611
Value of Casings	322,193	221,051	314,381	573,527	633,582	656,492	458,108	377,829	603,896	105,679
Percent of Total	72	74	77	74	73	71	71	76	61	47
Value of Inner Tubes	44,386	29,413	44,616	80,575	105,487	118,235	74,983	52,858	81,313	20,101
Percent of Total	10	10	11	10	12	13	12	11	8	9
Value of Solid and Other Tires	3,834	6,048	7,868	22,808	40,463	49,822	36,041	19,651	67,718	20,642
Percent of Total	1	2	2	3	5	5	6	4	7	9
Value of All Other Products	75,679	42,801	39,418	93,267	90,156	100,453	75,062	45,785	234,161	77,189
Percent of Total	17	14	10	13	10	11	11	9	24	35
Wholesale Price of All Tires — Composite Index	47	43	47	55	76	100	111	182	212	175

[a] Value data from the Census of Manufactures. Composite price index of all tires and tubes from the Bureau of Labor Statistics, Wholesale Prices.

[b] All value data are in thousands of dollars.

[c] These years are not exactly comparable to the others due to the fact that the industry was not separately classified by the Census until 1921. Tires, however, dominated the group in which they were classified (i.e. "Rubber Goods Not Elsewhere Specified" in 1914, and "Rubber Tires, Tubes, and Other Rubber Goods" in 1919) as is indicated by the proportions of non-tire products in these years. See pp. 29–30.

tically in the last decade, largely because of the almost continuous decline in prices. This trend is all the more remarkable because it has coincided with the large advance in tire durability just described. The details of the movements of values and prices in the industry are brought out more fully in Table V. The relative importance of the various products is also shown by major groups.

Since 1921 pneumatic tire casings have accounted for more that 70 percent of the value of products of the tire industry as classified by the Census. In the same period inner tubes have made up more than 10 percent of the total so that since 1921 pneumatic tires and tubes have constituted from 80 to 85 percent of the products of the industry by value. For a time solid tires threatened to become an important branch of the industry, reaching a total of 6 percent of the value of all products in 1923. The adoption of pneumatic tires for trucks proceeded at such a rapid pace, however, that solid tires soon dropped to insignificance.[24] A wide variety of other products, most of which are made of rubber, make up the other category of products of the tire industry. This group has constituted from 9 to 17 percent of the industry's total output since 1921, and prior to that date a larger proportion. The proportion of non-tire products has been increasing again since about 1931.[25]

[24] Nearly all trucks were equipped with solid tires until about 1915 but by 1920, only about 50 percent and by 1931 only 0.4 percent were so equipped. The total number of trucks in use increased so rapidly during this period, however, that the peak year of solid tire production was 1919, in which year nearly 1,500,000 were produced. In 1923 production again exceeded 1,000,000 but by 1929 the output had dropped to less than half this amount, and in 1935 only about 37,000 solid tires were produced. In value statistics of the industry various other specialized types of tires, including industrial truck, tractor, trailer, motorcycle and bicycle tires are frequently lumped with the solid tire group. It is only the growth in demand for some of these specialized lines which has kept this category from complete disappearance from the statistical reports of the industry in recent years. See Census of Manufactures and *Survey of Current Business.*

[25] Mechanical rubber goods are the most important other rubber products manufactured in the tire industry. They include rubber belting and hose, rubberized fabrics, rubber tubing, liner strips, washers, gaskets, engine

PRODUCTION 53

The principal markets for tires and tubes in the order of their importance are renewal sales, original equipment sales, and exports (See Table VI). Sales of tires to automobile producers for original equipment are almost exclusively in the hands of a few large producers. These orders are secured in large blocks, and in addition to serving as backlogs they are regarded as effective advertising for future replacement sales. As a consequence competition for these orders is very keen and the sales are usually made at a small profit and sometimes at an actual loss to the tire companies. The export business is also concentrated in the large producers since they are the only ones with sufficient resources to maintain the extensive organization required for this business. Although the United States has in the past sold as high as 30 percent of the total world exports of tires, the volume of such business has never been large as compared with domestic sales. At their peak in the late twenties exports accounted for about 5 or 6 percent of total sales, but in the more recent retreat of foreign trade in the face of the world-wide drive for national self-sufficiency, exports of tires have declined to about 2 per cent of total sales.[26]

From the above considerations it appears that the renewal tire market is by far the most important outlet for tire production since it is practically the sole market for the majority of

mountings, and many other similar products. The tire industry also manufactures tire repair materials, rubber boots, shoes, and rainwear, hard rubber products, and druggists and medical sundries. A complete list of products manufactured would run into the hundreds of articles.

[26] E. G. Holt, *International Shipments of Automobile Casings, Rubber Boots and Shoes, Rubber and Balata Belting, and Rubber Hose and Tubing, from Principal Manufacturing Countries*, United States Bureau of Foreign and Domestic Commerce, Circular Ru-3544, Washington, 1934, p. 1 (Mimeographed).

Automobile Facts and Figures, 1937 Edition, p. 38.

Imports of tires into the United States have always been utterly insignificant. In the year of their greatest volume (1924) they constituted less than 0.4 percent of the total sales in the United States and in recent years they have been only about 0.1 percent. Holt, Bureau of Foreign and Domestic Commerce, Circular Ru-3500, p. 14.

TABLE VI
ANALYSIS OF PRINCIPAL MARKETS FOR PNEUMATIC TIRES [a]
1910–1936

Year	Number of Tires (in thousands)				Renewal Sales per Car [b]	Automobile Registrations [c]
	Original Equipment	Renewal Sales	Exports	Inventory Change		
1910	724	1,525	51	+ 100	5.0	469
1911	797	2,031	72	+ 100	4.4	640
1912	1,424	2,971	105	+ 500	4.8	944
1913	1,846	4,022	132	—	4.5	1,258
1914	2,175	6,008	139	− 300	5.0	1,711
1915	3,584	7,871	495	+ 50	4.8	2,446
1916	6,139	10,782	644	+ 1,000	4.7	3,513
1917	7,086	16,754	496	+ 1,500	5.1	4,983
1918	4,046	20,494	460	− 1,000	4.4	6,147
1919	7,249	23,373	964	+ 1,250	4.1	7,565
1920	8,472	20,565	1,543	+ 1,820	2.9	9,232
1921	6,299	21,973	649	− 1,620	2.6	10,463
1922	10,113	28,477	1,438	+ 900	2.9	12,238
1923	16,409	27,796	1,489	− 270	2.4	15,092
1924	14,645	33,728	1,389	+ 1,240	2.4	17,595
1925	17,212	39,288	1,770	+ 540	2.3	19,937
1926	17,548	39,200	1,654	+ 1,740	2.0	22,001
1927	14,001	46,888	2,811	− 140	2.2	23,133
1928	18,019	52,303	2,689	+ 2,520	2.3	24,493
1929	22,067	45,471	2,979	− 750	1.9	26,501
1930	13,970	37,231	2,684	− 2,270	1.4	26,545
1931	10,429	37,983	1,959	− 980	1.4	25,833
1932	6,246	33,474	1,095	+ 100	1.3	24,115
1933	9,464	33,699	1,239	+ 990	1.4	23,844
1934	13,510	31,869	1,289	+ 570	1.3	24,952
1935	19,435	30,091	1,100	− 2,130	1.2	26,231
1936	21,921	31,353	1,077	+ 3,790	1.2	28,221

[a] E. G. Holt, Bureau of Foreign and Domestic Commerce, Circular Ru-3500, p. 14.

[b] Renewal sales per pneumatic tired car registered in the preceding year after allowance for scrappage of cars retired from use.

[c] *Automobile Facts and Figures*, 1937 Edition, p. 16.

tire producers and is the major source of the profits of the others. As shown in Table VI, renewal sales increased from about 1,500,000 in 1910 to more than 20,000,000 in 1920, and reached a peak of more than 52,000,000 in 1928, but have suffered a subsequent decline to only slightly more than 31,000,000 in 1936.[27] It is apparent that renewal sales have not kept pace with automobile registrations in recent years. This divergence is forcibly demonstrated by the decline of renewal sales per car, as shown in Table VI. In 1910 5.0 tires per car were sold but in 1920 only 2.9 were sold and by 1936 the figure had dropped to 1.2.[28] In view of the fact that automobile registrations themselves have been increasing much more slowly in recent years, the renewal sales market does not hold forth much promise of either immediate or long run expansion. Both the manufacturers and distributors of tires have been affected by this prospect. If in the future tires should be developed which will last as long as the average car, the tire industry may be reduced to the relatively minor role of a subsidiary adjunct to the automobile industry. Already during the past decade the narrowing market for renewal sales has been the scene of a struggle between small dealers and large scale distributors.

The recent conflict in distribution was originally precipitated by the series of contracts in force from 1926 to 1936 between the Goodyear Tire and Rubber Company and Sears, Roebuck and Company. Under these contracts Sears received Goodyear tires under private brand names at exceedingly favorable prices. Sears immediately built up a large tire business by underselling the tire dealers of the major companies, as well as its mail order and chain store competitors. Following in Sears' wake, other mail order houses, chain stores, and later oil companies secured

[27] It is estimated that 1937 and 1938 renewal tire sales were slightly below 30,000,000 per year. W. W. Leigh, " 1938 Prospects for the Independent Tire Dealer," *Tire Review*, January, 1938, pp. 9-11.

[28] Dr. Leigh estimated that in 1937, 1.05 renewal tires were sold per registered car and that in 1938 only about one replacement tire per registered car would be sold. *Ibid.*, p. 11.

similar contracts from the large tire producers. These developments aroused the ire of the manufacturers whose independent dealers were thus undersold. Firestone assumed the lead of this group and attempted to meet the new competition by drastic price slashes.

From 1926 to 1936 there was a succession of bitter price wars which made both tire manufacturing and tire distribution unprofitable for all but the most efficient operators, and forced a large number of manufacturers and independent dealers out of business. It was in this period also that Firestone initiated company-owned tire stores in a further attempt to defeat the private brand competition. The other large tire companies immediately followed by establishing stores of their own. In this period the consumers reaped large gains, being able at times to buy tires at cost and occasionally even below cost, but the producers suffered severely. Between 1926 and 1930, Sears' volume of tire sales increased more rapidly than that of any other distributor and by 1933 it was the largest retailer of tires in the United States.

The Federal Trade Commission instituted proceedings against the Goodyear-Sears contract in 1933 charging violation of the Clayton Act. In 1936 the Commission went to court to enforce its order to cease and desist, issued the same year. In the meantime Goodyear, following the passage of the Robinson-Patman Act, had abrogated the contract on the grounds that it did not wish to risk a violation of the new law, while maintaining that there had been no violation of the existing law. The issue still had to be fought out in the courts because Goodyear might be liable for triple damages to injured parties if the court held that the contract had violated the anti-trust laws. The lower court upheld the Commission but the Circuit Court of Appeals for the Sixth Circuit said that the Robinson-Patman Act and the cancellation of the contract had rendered the case moot. The commission went to the Supreme Court which remanded the case to the Circuit Court for rehearing. On February 16, 1939, the Circuit Court handed down a decision setting aside the Commission's order and upholding the agreement between Goodyear and Sears. The legal battle thus ended nearly thirteen years after the original Goodyear-Sears contract was made.[29]

During this period the mass distributors firmly established a position in tire retailing. The N. R. A. estimated that in 1922 independent dealers sold about 98 percent of all renewal sales, but that in 1934 their percentage of the total had been reduced to 58. In the latter year company stores and direct factory sales accounted for 10 percent, sales for spares on new cars, also largely factory controlled, made up an additional 6 percent, department store and mail order sales were 15 percent, and oil company sales were 10 percent of the total.[30] By 1937 the channels of distribution were becoming more stabilized or at least the shifts were occurring more gradually. In that year the proportions of tires sold by the various types of outlets were:[31]

Outlet	Percentage of Total Replacement Sales
Independent Dealers	52.0
Oil Companies	16.5
Chain Stores	14.7
Manufacturer Owned Stores	10.8
Mail Order Houses	4.0
Others [32]	2.0
Total	100.0

The differences between total sales and production are due to the increases or decreases in inventories as shown in Table

[29] Federal Trade Commission, *In the Matter of the Goodyear Tire and Rubber Company*, Docket 2116, Washington, Government Printing Office, 1936; John W. Crider, " F. T. C. Criticizes Anti-Trust Laws," Special Article in *The New York Times*, April 2, 1939, Section 1, p. 8; Hamilton *et al.*, *op. cit.*, pp. 83-116.

[30] Cross, Earseman and Lenaerts, *op. cit.*, p. 95. The remaining one percent was distributed by all other types of outlets.

[31] W. W. Leigh, "30,000,000 Tires Go to Market," *Tire Review*, June, 1938, pp. 12-15.

[32] Factory Direct 0.9, Department Stores 0.9, Cooperatives 0.2. Sales of spare tires for new cars are not included in these figures.

VI. Throughout most of the period since 1910 this item has been relatively unimportant, although in some years, notably 1928, 1930, 1935, and 1936 the changes in inventory have been considerable. In several years they have been as important as exports and in a few years more important. The total volume of inventories, as distinguished from annual changes, is fairly large in comparison with annual production. There has been some upward tendency in recent years partly because of a shift in inventories from dealers to manufacturers. The Rubber Manufacturers Association estimates manufacturers' inventories as of the end of the year as follows:[33]

Year	Pneumatic Casings	Inner Tubes
1927	10,312,000	13,692,000
1932	7,644,000	6,479,000
1936	11,114,000	10,985,000

Physical Productive Capacity

Another factor having an important bearing upon the volume of production of an industry is its physical capacity. In addition to setting a maximum limit upon output in the short run, the capacity also exerts a major influence upon costs. Business men try to operate their plants at the most economical percent of capacity (usually as near full utilization as possible) in order to distribute the burden over the maximum number of units.

It is somewhat hazardous to speak of the physical capacity of a given plant in exact terms and these difficulties are multiplied when an attempt is made to measure the capacity of an entire industry. The following discussion relates to the practical physical capacity of the tire industry, taking into account the bottlenecks in each plant and the stoppages required for repairs, maintenance, and contingencies. The estimates are given in terms of annual capacity, based upon data for daily maximum output assuming 250 days of operation per year.[34] It would

[33] *Statistical Bulletin of the International Rubber Regulation Committee*, London, August, 1937, p. 17.

[34] This is the basis of computing capacity which is recommended by the Rubber Manufacturers Association.

have been possible to have turned out a larger output than the capacity given in most of these years if the industry were working under the stress of emergency conditions, such as during a war, when cost becomes a subordinate consideration. Capacity to produce which can be used only at costs and prices materially above the normal range of the industry at the time has not been considered practical capacity. Insofar as possible the equipment relating to obsolete types of products has also been excluded from consideration. Since the estimates of capacity are in physical terms the ability of the market to absorb the total output has not been taken into consideration.

In Table VII are presented the annual physical capacity data in terms of number of tires and average percentage of capacity utilized for the period 1921 to 1938. Since the average weight and quality of tires have changed materially during this time it would have been better to have used data in terms of pounds but no such data are obtainable. Although some of the change in the weight of tires has been taken into account, it is probable that the basic data of individual plant capacities do not contain a full correction for this factor.

Although statistical data concerning potential tire production are not available prior to 1921, the history of the industry seems to indicate that prior to 1918 the annual output seldom dropped below 75 or 80 percent of capacity. The growth of demand was so rapid that plants and equipment barely kept pace with it. The war years stimulated the already rapidly growing demand, and in 1919 the pile-up in orders was so great that the industry apparently operated at a level which was in excess of its normal limit or that which could be sustained over a longer period. The middle of 1920, however, witnessed a sharp decline in demand, and in 1921 the tire industry ran at only 60 per cent of its possible production. Demand then recovered rapidly so that in 1922 it was operating at 88 percent of capacity, and in spite of the considerable increase in physical facilities from that year to 1925, in the latter year the industry was pressed to 91 percent of its capacity. By 1928 the industry had a capacity of

TABLE VII
PRACTICAL PHYSICAL CAPACITY OF THE TIRE INDUSTRY
1921-1938

Year	Practical Capacity [a] (Thousands of Tires)	Percent of Capacity Used [b]
1921	45,600	60
1922	46,500	88
1923	61,800	74
1924	62,000	82
1925	64,700	91
1926	68,600	89
1927	74,500	85
1928	86,700	87
1929	91,600	76
1930	90,000	57
1931	86,000	57
1932	82,000	49
1933	77,000	59
1934	72,000	66
1935	68,000	71
1936	70,000	83
1937	75,000	74
1938	77,000	53

[a] Data for the period 1921 to 1929 are the estimates of Leonard Smith of the Commercial Research Department of the United States Rubber Company as quoted by E. G. Nourse in *America's Capacity to Consume*, The Brookings Institution, Washington, 1934, p. 583. Data for the period 1930 to 1938 are estimates made by the writer based upon Cross, Earseman and Lenaerts, *Rubber Industry Study, op. cit.*, pp. 29-34, unpublished data in the N.R.A. files, and personal correspondence and interviews with executives of the Rubber Manufacturers Association and of several individual tire companies.

[b] Production (Table I, Column 1) divided by practical capacity (Column 1 of this table).

86,700,000 tires as compared with only 45,600,000 in 1921, yet in 1928 it was working at 87 percent of its maximum. The capacity was still further increased in 1929 but production fell in the closing months so that for the full year the industry averaged only 76 percent of full utilization. The 1929 capacity was, however, slightly more than double that available in 1921.

Between 1929 and 1935 the capacity of the industry declined from 91,600,000 to only 68,000,000 tires per year. The main cause of this development was the failure to replace much of the equipment which wore out or became obsolete. Another contributing factor was the manufacture of a larger proportion of non-tire products in tire factories. Since 1935, however, the capacity of the industry has been increasing once more, and it is estimated that in 1938 the 45 plants in the industry had an annual capacity of about 77,000,000 tires.

The drop in production from 1929 to 1932 was much more rapid than the drop in capacity so that in the latter year the industry operated at an average of only 49 percent of its capacity. In the years 1933-1935, capacity was still further reduced while production was increasing, and by 1935 the industry was operating at 71 percent of its capacity. In 1936 production increased more rapidly than capacity and the industry operated at the highest percent of its capacity since 1928, but production receded once more in 1937 and 1938 while the effects of the new construction in the industry were beginning to be realized in the form of greater capacity. The drastic decline in production in 1938 forced the percent of capacity used to 53 percent, the lowest figure since 1932. The industry suffered seriously from overcapacity during the period 1930-1934, and although it is not substantially overbuilt at the present time, it is in danger of becoming so if it undergoes a wave of new building which is not counterbalanced by an increase in demand or a decrease in the capacity of the existing facilities. Since the new facilities are nearly always the most efficient, however, they usually tend to make some of the existing capacity obsolete.

CHAPTER IV

THE LABOR SUPPLY

Employment and Man-Hours

The output of the tire industry requires a large labor supply. The number of wage-earners employed in the manufacture of tires, estimated to have been between 10,000 and 15,000 in 1909, rose to nearly 90,000 in 1919 but dropped to less than 50,000 in 1921. In 1929 about 72,000 were so engaged, but by 1937 the level had declined again to about 50,000.[1] The number of wage-earners in establishments the major product of which is tires, a basis on which we have more accurate data, increased from 50,000 in 1914 to nearly 120,000 in 1919. In the post-war depression of 1921 employment dropped to 55,000, but it increased again to nearly 74,000 in 1923 and more than 83,000 in 1929. The depression which began in that year caused a decline to only slightly above 45,000 in 1932, but in 1937 employment had risen again to 66,000. A sharp but relatively short-lived drop occurred again in 1938. The course of employment in the intervening years is shown in Table VIII.

Such measures, relating to the average number of wage-earners at work during the years given, reflect man-years of employment. This unit is not entirely satisfactory for comparisons extending over considerable periods because the average number of hours included in a man-year has declined throughout the period. Table VIII indicates that average hours per week have dropped from 49.1 in 1914 to 44.8 in 1929 and to 31.7 in 1937. Total man-hours of labor expended in the tire

[1] These estimates of the number of wage-earners actually employed solely in manufacturing tires were made by the writer on the basis of Census of Manufactures and Bureau of Labor Statistics data for the industry by deducting from total employment the estimated number of wage-earners employed on non-tire products in these establishments. These latter deductions were based upon the relative value of tire and non-tire products in these establishments for the given years. Although these estimates have rather wide margins of error, particularly in the earlier years, they are believed to give a fair indication of the trends over the interval.

TABLE VIII

EMPLOYMENT IN THE TIRE INDUSTRY FOR SELECTED YEARS, 1914–1938

Indexes: 1925 = 100

Year	Wage-Earners [a] Number	Index	Average Weekly Hours [b] Number	Index	Man-Hours [c] Number (Thousands)	Index
1914	50,220	62	49.1	110	123,290	68
1919	119,848	147	—	—	—	—
1921	55,496	68	43.8	98	122,536	67
1923	73,963	91	44.8	101	164,301	91
1924	71,537	88	44.1	99	158,035	87
1925	81,640	100	44.5	100	181,649	100
1926	79,788	98	45.0	101	179,833	99
1927	78,256	96	45.7	103	178,017	98
1928	83,194	102	46.1	104	192,548	106
1929	83,263	102	44.8	101	186,509	103
1930	59,803	73	41.4	93	123,521	68
1931	49,159	60	38.8	87	94,457	52
1932	45,269	55	32.5	73	78,089	43
1933	52,976	65	31.6	71	83,559	46
1934	61,241	75	30.7	70	94,457	52
1935	57,128	70	32.3	73	92,262	51
1936	61,696	76	35.4	80	108,989	60
1937	66,047	81	31.7	71	104,684	58
1938	47,700	58	29.6	67	72,008	40

[a] Census of Manufactures, 1914, 1919 and 1921. Bureau of Labor Statistics, *Employment and Payrolls*, for subsequent years. The 1914 and 1919 data are not exactly comparable with subsequent years because of the change in the Census classification of the industry in 1921. All data are comparable with the production figures given in Chapter II, however. The Bureau of Labor Statistics data apply to the industry as defined by the Census since 1921.

[b] 1914 to 1931 data from The National Industrial Conference Board. (See M. Ada Beney, *Wages, Hours and Employment in the United States, 1914-1936*, New York, 1936, pp. 148-151); data since 1932 from the Bureau of Labor Statistics, *Employment and Payrolls*.

[c] Number of wage-earners multiplied by average hours per week.

industry (as also shown in Table VIII) have been computed by multiplying the average number of workers employed by the average weekly hours and converting to the annual basis. The man-hour measure corresponds fairly closely with that for man-years until 1929. Beginning in 1930, however, average hours per week dropped so rapidly that in 1937 they were approximately 13 hours or 29 percent less than in 1929, and as a consequence the decline in man-hours was considerably greater than that in man-years during this time. The man-hour index on a 1925 base dropped from 103 in 1929 to 43 in 1932, and in 1937 had risen only as high as 58. The drop in the man-year index, also on a 1925 base, was from 102 in 1929 to 55 in 1932 and by 1937 it had risen to 81. In 1938 man-hours reached a point below that of any year since 1914. The man-hour is a better measure of the labor requirements of the industry but the average number of workers employed is also significant because it indicates the number of individuals involved. These two measures of employment will therefore be used concurrently in the remainder of the study.

CHARACTERISTICS OF THE LABOR FORCE

During the period of rapid expansion the industry faced the problem of recruiting an adequate labor force. Many of the operations involved in the manufacture of tires require a very considerable amount of skill as well as unusual physical strength and since not many qualified workers were available the industry largely trained its own workers. The fact that laborers trained in the rubber industry were available in Akron after the location of the Goodrich Company there in 1870 was a factor in the concentration of the tire industry in that city.[2] The Akron plants being new in the industry, were in a favorable position to concentrate on tires, the then rapidly developing product which has since overshadowed the remainder of the industry. This city thus became the tire producing center of the country and its labor supply the major part of that of the tire industry. Native labor has predominated throughout

[2] Frazier and Doriot, *op. cit.*, p. 106. These authors suggest that this explanation may also be applicable to the concentration of the Pacific Coast production in Los Angeles in recent years. This factor is further discussed in Chapter VIII below.

the period and, although the proportion of foreign born was once as high as 20 percent, it has since declined to a much smaller figure.

The available evidence indicates that the earliest workers in the Akron plants were native Americans, many of whom had but recently arrived in Ohio from New England, but with a liberal intermixture of Irish and German immigrants. The expanding industry drew in workers from the nearby small towns and farms in Ohio, and was already recruiting many workers from Pennsylvania, West Virginia, and Kentucky in the prewar years. Many immigrants, particularly those from the Slavic countries of central and eastern Europe, found jobs in the industry in large numbers also during these years. The war-time expansion drew a large number of workers from the deep South as well as an increase in the number from the neighboring states. A considerable sprinkling of Negroes also was recruited from the South and brought into the industry chiefly during the war period.

Almost since the beginning of the industry a considerable number of women have been employed in tire factories, although in the early days the majority of them worked on products other than tires. The percentage of women employed was declining prior to the war as the proportion of tires increased relative to total output. Thus in 1909 the Census of Manufactures showed that 17.1 percent of the wage-earners in " Rubber Goods Other Than Boots and Shoes and Mechanical Goods " were women, whereas in 1914 women were only 13.1 percent of the comparable category.[3] The relative scarcity of labor in the war years led to a large expansion in the number of women employed, but the proportion of women declined slightly to 12.6 percent in 1919. Between 1928 and 1935 the proportion of women wage-earners of the Goodyear Company's

[3] In 1914 the Census changed the name of this group to " Rubber Goods not Elsewhere Specified."

Akron factories ranged between 10 and 13 percent.[4] The N. R. A. estimated that about 16 percent of the wage earners of the tire industry were women in 1929, and that the proportion had increased slightly in the depression but was declining again in the recovery, so that it is probable that the proportion of women wage-earners remains about 15 percent at the present time.[5] The chief occupations in which women are engaged are in the production of inner tubes, motorcycle and bicycle tires, and the light non-tire products, but considerable numbers of them are also employed in some of the lighter occupations in tire casing production.

Child labor, defined as the labor of persons under 16 years of age, has constituted less than one percent of the total employment in the industry since 1914 and has been practically non-existent since the World War.[6]

Workers have enjoyed somewhat higher earnings and shorter hours than in most other American industries and, although working conditions have not always been good, marked improvements have been made in the last twenty years. Industrial accidents, which were all too frequent in the early days, have been greatly reduced, and the use of materials which are poisonous or harmful to the workers has been discontinued or surrounded with safeguards.[7]

---

[4] Fred C. Croxton, John A. Lapp, and Hugh S. Hanna, *Findings and Recommendations of the Fact Finding Board Appointed by the Secretary of Labor, November 15, 1935*, Washington, December 16, 1935, p. 33 (Mimeographed).

[5] Kress and Pearce, *op. cit.*, p. 25.

[6] Less than six-tenths of one percent of the employees were under 16 according to the 1919 Census of Manufactures (Rubber Industry, Table 3). Since that time the Census has ceased collecting these data.

[7] An indication of the accident hazard in the tire industry as compared with other industries may be obtained by an examination of Workmen's Compensation premiums by industries. These rates are graduated according to the number and seriousness of the accidents experienced. In 1936 the rates of a selected list of major industries in Ohio were:

Industry	Premium per $100 Payroll
Clothing Manufacturing	$0.30
Printing	0.35
TIRE MANUFACTURING	0.90
Automobile Manufacturing	1.00
Iron and Steel Manufacturing	1.40
Foundries	1.80
Carpentry	3.50
Coal Mining	7.00

On the whole the tire industry makes a favorable showing in safety. See Mary J. Drucker, *The Rubber Industry in Ohio*, National Youth Administration in Ohio, Occupational Study No. 1, Columbus, Ohio, 1937, p. 50.

CHAPTER V

THE PRODUCTIVITY OF LABOR

TOTAL MANUFACTURING LABOR PRODUCTIVITY—DIRECT AND INDIRECT

THE relationships between production and employment may be interpreted more precisely by means of techniques involving the measurement of the average productivity of labor in the industry. The physical volume of production of the tire industry has been measured in Chapter III in terms of the equivalent number of tire casings produced, the amount of rubber consumed in their manufacture, and the number of tire-miles produced. The quantity of labor employed in the manufacture of tires has been measured in Chapter IV in terms of man-years and man-hours. These measures of production and employment have been used to derive six types of measurements of the productivity of labor, namely: (1) tires per man-year; (2) tires per man-hour; (3) pounds per man-year; (4) pounds per man-hour; (5) tire-miles per man-year; and (6) tire-miles per man-hour. These data are shown in Table IX. Six series have been used because each has its place in the succeeding analysis. Thus the first two series are most nearly comparable with productivity studies in other industries, the third and fourth are more suitable for cyclical analysis, and the last two series make possible fuller comparisons with labor costs and tire prices. In addition to the above examples each series has several other uses.

The simplest but least exact of the measures of productivity is in terms of tires per man-year. Nearly twice as many tires per man per year were made in 1921 as in 1914, and in 1925 more than three times as many tires were made per man-year as in 1914. Since 1925 the movement of tires per man-year has fluctuated rather widely, but only a very moderate increase in productivity has been shown. There was an unsteady upward movement from 1925 until 1931, and in the latter year 29 per-

TABLE IX

TOTAL LABOR PRODUCTIVITY PER WORKER AND PER MAN-HOUR IN TIRE MANUFACTURING FOR SELECTED YEARS, 1914–1937

Indexes: 1925 = 100

Year	(1) Tires per Man-Year [a]	(2) Tires per Man-Hour [b]	(3) Pounds per Man-Year [c]	(4) Pounds per Man-Hour [d]	(5) Tire-Miles per Man-Year [e]	(6) Tire-Miles per Man-Hour [f]
1914	33	30	—	—	11	10
1919	44	—	—	—	—	—
1921	64	65	57	58	31	31
1923	85	85	73	73	—	—
1924	98	99	94	95	—	—
1925	100	100	100	100	100	100
1926	103	102	94	93	—	—
1927	109	107	95	93	—	—
1928	122	117	108	104	—	—
1929	109	111	113	112	162	161
1930	113	121	123	132	—	—
1931	129	148	138	160	—	—
1932	116	150	135	185	218	300
1933	114	161	137	193	—	—
1934	104	161	131	188	—	—
1935	117	160	151	208	224	308
1936	128	162	166	209	—	—
1937	115	160	162	227	233	325

[a] Equivalent number of tire casings (Table I, Column 3) divided by average annual number of wage-earners employed (Table VIII, Column 2).

[b] Equivalent number of tire casings divided by man-hours (Table VIII, Column 5).

[c] Pounds of crude rubber consumed (Table I, Column 4), divided by average annual number of wage-earners employed.

[d] Pounds of crude rubber consumed divided by man-hours.

[e] Tire-miles produced divided by average annual number of wage-earners employed. Tire-miles produced computed by multiplying index of equivalent number of tire casings (Table I, Column 3) by index of average mileage per casing (Table II, Column 1) converted into index numbers on the base 1925 = 100. The average mileage for 1915 was used for 1914.

[f] Tire-miles divided by man-hours.

cent more tires were produced per man-year than in 1925. The next four years witnessed a decline, however, and in 1934 the level was only 4 percent above 1925. The next two years saw a sharp rebound and in 1936 the level was 28 percent above 1925, but the following year saw a drop again so that in 1937 tires per man-year were only 15 percent above the level of 1925. Tires per man-year is an inadequate basis for the measurement of productivity, however, because both of the units involved in the measure, the tire and the man-year, have undergone substantial changes during the period. On the one hand the tire of 1937 is a quite different product from that of 1914 because the 1937 tire will wear nearly six times as long as the one made in 1914. In similar fashion the man-year of 1937 is different from that of 1914 because in 1937 the average wage-earner in the industry worked about 35 percent fewer hours in the course of the year than did the average worker in 1914. The latter difficulty has been removed from the productivity measurements by the use of the tires per man-hour index.

Tires per man-hour more than doubled between 1914 and 1921 and more than tripled between 1914 and 1925. The movement since 1925, although not without occasional slight falls, has been more consistently upward in tires per man-hour than in tires per man-year, and the former index has shown a substantially greater net increase in the period since 1925. Between 1925 and 1933 tires per man-hour increased 61 percent, but since the latter date the series has fluctuated about a nearly constant level.

As has been indicated in Chapter III, pounds of rubber consumed represents a better measure of physical production than the number of tires. Although adequate data on this basis are not available prior to 1921, the measurement of productivity by the use of this production series is more accurate than in terms of number of tires. More than 75 percent more pounds of rubber were made into tires per man-year in 1925 than in 1921. Since 1925 the increase in pounds per man-year has also been nearly continuous, and by 1937 productivity in pounds per man-year had increased by no less than 62 percent over 1925.

A better measure of the productivity of labor in the industry is in terms of pounds of rubber consumed per man-hour. The results obtained by the use of this measure are very nearly the same as those in terms of pounds per man-year through the year 1929. Thus pounds per man-hour increased 73 percent between 1921 and 1925 and 12 percent between 1925 and 1929, while the corresponding rates of increase in pounds per man-year were 75 percent and 13 percent, respectively, for these periods. Productivity in pounds per man-hour increased at a much more rapid rate than pounds per man-year after 1929, however. Thus, in 1937 more than twice as many pounds of rubber were made into tires per man-hour than in 1925, whereas the increase in pounds per man-year was only 43 percent in this period.

As has been pointed out in Chapter III, however, the measurement of the physical volume of production in terms of rubber consumed is by no means a full correction for the changes in the quality of tires. To the extent that production in pounds understates the physical volume of production, the increase in the productivity of labor over the period is also understated. In order to indicate the magnitude of this understatement of the productivity increase, productivity has been measured in terms of tire-miles per man-year and tire-miles per man-hour. On this basis tire-miles per man-year have increased no less than nine-fold between 1914 and 1925 and 133 percent between 1925 and 1937. Tire-miles per man-hour have increased at the even more rapid rate of 10-fold between 1914 and 1925 and 225 percent between 1925 and 1937. Between 1914 and 1937 tire-miles per man-year have increased more than 20-fold and tire-miles per man-hour have increased more than 30-fold. These data as well as the changes in some of the intervening years are shown in Table IX.

DIRECT LABOR PRODUCTIVITY

The above productivity measures take into consideration all wage-earners in the tire manufacturing industry, whether or not their work is direct production labor. This procedure has been used in the productivity analysis because it is believed that the results have a somewhat broader significance than those which are to be obtained by comparisons of changing productivity in terms of direct labor alone. Since a considerable proportion of the labor in tire factories is indirect it is clear that the productivity measures to be obtained from direct labor alone are considerably higher than those given above.

Mr. Boris Stern of the United States Bureau of Labor Statistics has made a detailed study of the productivity of direct manufacturing labor in six large tire plants for the period 1922 to 1931.[1] These six plants produced 45 percent of the total product of the industry in 1922, but because this was a period of concentration, the proportion of total production included in these six plants increased steadily and stood at approximately 60 percent in 1931. The study took the form of an intensive field investigation in which the number of tires, the number of pounds of rubber consumed, the number of direct labor employees, and the number of man-hours of direct labor were obtained for each plant. All data secured relate to the production of pneumatic tire casings, so that no conversions of other products into equivalents were required. This method has the obvious advantage of eliminating all inaccuracies of conversion. The man-year basis was not used, so that productivity is measured in terms of tires per man-hour and pounds per man-hour only. The results obtained by Stern are shown in Table X.

A comparison of the indexes of direct labor productivity per man-hour, as shown by Stern, with those based upon all

[1] In defining the scope of his survey Stern states: "The term 'man-hours,' as used in this survey, covers direct productive labor only, that is, the labor directly and intimately involved in the process of production. Warehousemen, laboratory workers, foremen, checkers, timekeepers, etc., whose services are not directly involved in the process of tire making, are therefore not included in the figures for the man-hours used in this survey." *Op. cit.*, p. 5.

labor in the industry shown in Table IX indicates that between 1914 and 1925 the productivity of direct labor increased at about the same rate as the productivity of all labor. It appears, however, that since 1925 direct labor productivity has been increasing faster than indirect and consequently also faster than the productivity of labor as a whole.[2]

Stern's study also reveals large differences in productivity between separate plants. Thus, in 1922 the plant with the lowest productivity showed a production of .42 tires and 7.38 pounds per direct labor man-hour, while the plant with the highest pro-

TABLE X

Direct Labor Productivity Per Man-Hour in Six Large Tire Plants for Selected Years, 1914-1931 [a]

Year	Actual Output per Man-Hour		Indexes: 1925 = 100 Output per Man-Hour	
	Tires	Pounds	Tires	Pounds
1914	0.25	4.50	31	32
1922	0.70	11.28	88	82
1925	0.80	13.77	100	100
1929	1.07	22.80	134	166
1931	1.37	30.67	171	223

[a] Stern, *op. cit.*, pp. 7-9.

ductivity showed a production of 1.64 tires and 18.98 pounds per direct labor man-hour. In 1931 the plant with the lowest productivity produced .60 tires and 19.40 pounds per direct labor man-hour, while the plant with the highest productivity produced 2.44 tires and 39.22 pounds per direct labor man-hour.[3] The plant with the highest productivity in 1922 was also highest in 1931, but the plant with the lowest productivity in 1922 had moved up in rating by 1931. During the interval 1922 to 1931, the range between the lowest and highest productivity

---

[2] This conclusion is confirmed by Messrs. Kress and Pearce of the N.R.A. Research and Planning Division in *Material Bearing on the Rubber Tire Industry, op. cit.*, pp. 31-38.

[3] *Op. cit.*, pp. 13-14.

of these plants increased slightly in terms of tires per man-hour, but decreased in terms of pounds per man-hour. As might have been expected, the largest increases in productivity in the period came in plants with relatively low productivity at the beginning of the period.[4]

Direct Labor Productivity by Departments of Tire Manufacturing

In the measurement of total labor productivity, including both direct and indirect labor, it is practically impossible to show the differences between each of the various operations with regard to productivity increase. Stern has, however, succeeded in doing this with regard to direct labor productivity. The classification of the direct operations into departments involved considerable difficulties, not only because of the variations in methods and classification between plants, but even more because of the many changes in operations and record forms during the period. Stern, therefore, divided the entire process of tire manufacturing into three major parts, namely: (1) crude rubber preparation; (2) stock preparation and tire building; and (3) curing, finishing and inspection.

Crude rubber preparation consists of several operations involving the cutting, washing, grinding and breaking down the crude rubber in order to remove impurities and to facilitate the handling of the rubber in later processes. Also included in this category are the operations of milling the rubber, compounding it with various chemical ingredients, and thoroughly mixing the resulting mass. Calendering the rubber, that is, sheeting it in combination with cord materials so as to impregnate the cord with rubber, is the last stage in this process.

The next division of manufacture consists of stock preparation and tire ("carcass") building. The calendered sheet stock is cut into strips for tire plies ready for the tire builders.

[4] One plant with relatively low productivity experienced no increase in tires per man-hour during the period, but productivity in pounds per man-hour in this same plant nearly doubled during that time.

The wire beads to go around the inside edges of the tires, the rubber tread, and the sidewalls are made in this department, as well as the extra strips of rubber or rubberized fabric (breakers, chafers, cushions and inserts) which also go into the tires. The last operation in this department is the construction of the body, or carcass, of the tire by putting the various parts together.[5]

The final set of operations includes curing, finishing and inspecting the tires. Curing consists of subjecting the tire to heat under pressure, thus completing the process of vulcanization started in the mixing operations of the crude rubber preparation. The average cure now lasts less than one hour, although this depends both upon the size of the tire and the accelerators used in compounding. Finishing consists of trimming away the overflow rubber left by the curing mold, washing and painting the tire. Inspection consists of making a thorough examination of the tire, inside and outside, in order to detect any possible flaws.

Stern's data on direct labor productivity per man-hour in these three divisions of the manufacturing operations are shown in Table XI. These results indicate that productivity in crude rubber preparation advanced at a slower rate than in all occupations, that productivity advanced most rapidly in stock preparation and tire building, and that the advance in productivity in curing, finishing and inspection was slightly less than that of all occupations. Stern notes that the most important major change in the manufacture of pneumatic tires was the shift from the core to the drum process of tire building. This is an important reason for the greater growth of productivity in this department. He also states that there have been exceptionally large reductions in labor requirements in bead-making and in

[5] The process of shaping tires built by the flat drum method (by which nearly all passenger car tires are now built) is performed in the curing department. Since, however, the shaping of the tires was a part of the process of tire building by the former method, this operation is included in the stock preparation and tire building department for the sake of making the data comparable over the period. Stern, *op. cit.*, p. 52.

some of the other steps in stock preparation.⁶ Stern concludes, however, that the greater part of the productivity increase in tire manufacturing in this period was due to the cumulative effect of a large number of evolutionary small changes, rather than to any particular major change in technology.⁷

TABLE XI

DIRECT LABOR PRODUCTIVITY BY DEPARTMENTS IN SIX LARGE TIRE PLANTS FOR SELECTED YEARS, 1922–1931 [a]

Output per Man-Hour. Indexes: 1925 = 100

	1931	1929	1925	1922
All Departments				
Tires per Man-Hour	171	134	100	88
Pounds per Man-Hour	223	166	100	82
Crude Rubber Preparation				
Tires per Man-Hour	147	115	100	86
Pounds per Man-Hour	177	135	100	79
Stock Preparation and Tire Building				
Tires per Man-Hour	181	143	100	94
Pounds per Man-Hour	235	175	100	87
Curing, Finishing and Inspection				
Tires per Man-Hour	164	129	100	85
Pounds per Man-Hour	212	158	100	79

[a] Stern, *op. cit.*, pp. 44, 55, and 63.

It is highly probable that much greater differentials in productivity increase would appear between separate operations if it had been possible to classify them more minutely. Thus, some operations appear to have remained virtually unchanged with regard to productivity, whereas others have been largely, and in some cases completely, eliminated by improvements in the productive process.

6 *Ibid.*, pp. 24-25 and 50.
7 *Ibid.* See Chapter VI.

Productivity in Value Terms

The discussion of productivity has been carried on entirely in physical terms thus far, leaving out the complicating factor of the values placed upon the products by the market. This factor appears in Table XII, in which the value of products per man-year and value added by manufacture per man-year and per man-hour are shown. Comparison has also been made between the tire industry and all manufacturing industry in these respects. The series relating to value added by manufacture are most significant, since the cost of raw materials is thus eliminated from the value data. It is from the value added by manufacture that the returns of all of the factors of production involved in the manufacturing operations must be taken.

It is clear that value productivity rose much less rapidly than physical productivity. Physical productivity between 1914 and

TABLE XII

Value Productivity in the Tire Industry and in All Manufacturing Industry for Selected Years, 1914–1935 [a]

(Dollars)

Year	Value of Products per Man-Year		Value Added by Manufacture per Man-Year		Value Added by Manufacture per Man-Hour	
	Tires	All Mfg.	Tires	All Mfg.	Tires	All Mfg.
1914	4,453	3,446	1,941	1,404	.76	.52
1919	8,236	6,862	3,850	2,753	1.77	1.10
1921	8,940	6,284	3,686	2,637	1.62	1.11
1923	8,710	6,904	3,773	2,948	1.62	1.15
1925	11,330	7,477	4,472	3,194	1.93	1.27
1927	11,113	7,511	4,734	3,268	2.01	1.33
1929	9,250	7,930	4,090	3,603	1.76	1.43
1931	8,265	6,307	4,733	2,975	2.39	1.41
1933	5,650	5,178	2,758	2,401	1.84	1.27
1935	7,809	—	3,161	—	1.88	—

[a] Census of Manufactures for value of products, value added by manufacture, and number of wage-earners. Man-hours, computed by the writer using Census data for number of workers and National Industrial Conference Board and Bureau of Labor Statistics average hours per week. See Footnotes to Table VIII.

1935 increased more than 250 percent in terms of tires per man-year and approximately 3,150 percent in terms of tire-miles per man-hour. In the same period the value of tires per man-year increased only 75 percent, the value added by manufacture per man-year increased only 63 percent, and the value added by manufacture per man-hour increased only 148 percent. A lower rate of increase in value productivity than in physical productivity was, of course, to have been expected in view of the movements of prices in the industry. It is notable, however, that value productivity increased rapidly from 1914 to about 1927, and that it has declined since that time. The peak in value of products in the year 1925 and its subsequent decline may be explained in part by the fall in raw materials prices since that time. Value added by manufacture per man-year, however, has declined since 1927 and value added per man-hour was lower in 1935 than in 1931. Value productivity rose as long as the increase in physical productivity proceeded at a more rapid rate than the fall of prices. The fall in value productivity in recent years is principally due to the fact that physical productivity has levelled off while prices have continued to decline.

It may be seen that the trends of value productivity in manufacturing industry taken as a whole are very similar to those of the tire industry. On the average between 1914 and 1935, the tire industry has shown about a 30 percent higher value productivity per man-year than all manufacturing industry, both on a value of products and on a value added basis. In terms of value added per man-hour, the differential in favor of the tire industry appears to have been about 40 percent.

The Physical Productivity of the Tire Industry Compared with the Rest of the Economy

The physical productivity of labor in the tire manufacturing industry has advanced at a markedly more rapid rate than that of the economy as a whole. Measurements of the physical productivity of labor in the national economy as a whole on a per man-year basis and of the productivity of labor per man-year and per man-hour in all manufacturing industries as a group are shown in Table XIII. The productivity of all employed labor in the country has been measured by the National Resources Committee by dividing an index of the total physical volume of production by an index of total employment, both series being on an annual basis. These data show an increase of man-year productivity of 15 percent from 1921 to 1925 and a further increase of 9 percent from the latter date to 1935. By contrast, the tire industry on the most conservative basis of productivity measurement, tires per man-year, exhibited a 56 percent increase from 1921 to 1925 and a 17 percent increase between 1925 and 1935. In terms of the more accurate measure of pounds of crude rubber consumed per man-year, the tire industry experienced a gain of 75 percent from 1921 to 1925 and 51 percent more between 1925 and 1935.

The productivity of labor in the entire economy, although an extremely useful concept in the interpretation of economic progress, cannot be said to be a very exact one because of the relative paucity of necessary data and the number of assumptions upon which the final measures rest.[8] Moreover, the movements of the final productivity measures reflect to a considerable degree shifts in the relative importance of various industries rather than real changes in productivity in the separate industries.[9] Manufacturing industry was the group in which the productivity of labor was increasing most rapidly. Since the tire industry belongs in this group the relative growth of productivity in tire manufacturing and in the whole manufacturing group forms a logical basis of comparison. As may be seen from Table XIII, Columns 2 and 3, productivity in all manufacturing on a man-year basis increased 28 percent from 1914

---

[8] Weintraub and Posner, *op. cit.*

[9] *Ibid.*, pp. 27-30. The authors especially stress the important shift from the basic industries (agriculture, manufacturing, construction, transportation, and communication) to the service industries. The productivity of labor in the latter industries was apparently not only at a considerably lower level but was increasing at a slower rate.

TABLE XIII

PHYSICAL PRODUCTIVITY IN THE NATIONAL ECONOMY AND IN THE MANUFACTURING INDUSTRIES GROUP IN THE UNITED STATES FOR SELECTED YEARS, 1914–1935

Indexes: 1925 = 100

Year	Total Economy Production per Man-Year [a]	Manufacturing Industries Production per Man-Year [b]	Manufacturing Industries Production per Man-Hour [c]
1914	—	78	72
1919	—	80	71
1920	79	—	—
1921	87	82	81
1923	94	94	91
1925	100	100	100
1927	97	105	103
1929	99	114	111
1931	94	110	124
1933	106	105	133
1935	109	108	142

[a] David Weintraub, and Harold L. Posner, *Unemployment and Increasing Productivity*, The National Research Project of The Works Progress Administration, Washington, 1937, p. 20. This bulletin is a reprint of Section V of The National Resources Committee Report, *Technological Trends and Their Social Implications*, Washington, 1937. The base of the indexes has been shifted by the writer.

[b] Data for the period 1914–1929 are from Frederick C. Mills, *Economic Tendencies in the United States*, National Bureau of Economic Research, New York, 1932, pp. 26, 192, and 290; 1931 and 1933 data are from Bulletins of the National Bureau of Economic Research, No. 45, February 20, 1933, p. 4 by Mills, and No. 58, November 15, 1935, p. 9 by Charles A. Bliss; the 1935 data, together with certain revisions of Bulletin No. 58, were given to the writer by Mr. Bliss in a personal interview. The basic data for the entire series are from the Census of Manufactures. The base of the indexes has been shifted by the writer.

[c] Data for the period 1914–1929 were computed by the writer from production and man-hour data obtained from The National Industrial Conference Board, *Machinery, Employment and Purchasing Power*, New York, 1935, p. 55. Data for the period 1931–1935 were obtained from the same sources as the man-year productivity data for these years. The base of the indexes has been shifted by the writer.

to 1925 and 8 percent between 1925 and 1935, while productivity per man-hour increased by 39 and 42 percent for these two respective periods. Thus the net increase in productivity per man-year for all manufacturing industries from

1914 to 1935 was 38 percent and the increase per man-hour was 97 percent. The corresponding increases in the tire industry in the same period were 255 percent in tires per man-year, and 433 percent in tires per man-hour.[10]

The series representing productivity in the entire economy and in all manufacturing industries are somewhat understated because they do not include allowances for the improvements in the quality of many products (e. g., automobiles). Moreover, these series are general averages including industries with decreasing or slowly increasing productivity as well as industries in which productivity is advancing rapidly. Thus it is not surprising to find that the tire industry showed a substantially greater rate of productivity increase than these groups. However, the United States Bureau of Labor Statistics, in a study of the productivity of labor in eleven manufacturing industries selected to include industries showing the most rapid increases, found that the tire industry was at the top of the list for the period 1914 to 1927.[11] Likewise the 1936 *Handbook of Labor*

---

[10] Lack of data makes it impossible to compute all manufacturing productivity series which would be comparable to tire-miles per man-year or per man-hour or to similar series for rubber consumed.

[11] "Productivity of Labor in Eleven Manufacturing Industries," *Monthly Labor Review*, Volume XXX, March, 1930, pp. 501-517. This study revealed the following results:

Industry	Percent of Increase in Man-Hour Productivity 1914–1927
Rubber Tires	292
Automobiles	178
Petroleum Refining	82
Flour Milling	59
Iron and Steel	55
Cement	54
Leather Tanning	41
Paper and Pulp	40
Cane Sugar Refining	33
Slaughtering and Meat Packing	26
Boots and Shoes	24

The productivity increase in all manufacturing industries as a whole in this

82   THE TIRE MANUFACTURING INDUSTRY

Statistics, although showing that some divisions of other industries have experienced a more rapid rise in productivity than the tire industry, shows only one industry (electric lamps) with a greater productivity increase than the tire industry from 1914 to 1931.[12]

Productivity changes in 17 industries are summarized in this bulletin. It is, of course, possible that some industries which have not been studied would show greater productivity increases than tires in this period. The rate of productivity increase in the tire industry has been greatly reduced since 1931, however, and since that date several industries have shown greater advances in productivity than tires.[13]

Nevertheless, taking the period since 1914 as a whole, the tire industry has been outstanding in this regard. The chapters to follow are concerned with some of the many ramifications of this development. In the analysis of the factors involved the investigation will be concerned in part with particular situations

period was 35 percent, so that all but three of these industries were above the general average.

Although this study does not specifically state the unit of production used in computing productivity in the tire industry, it appears from the results that the number of tires was used. As has been repeatedly pointed out above, this method gives the most conservative results of the three measures on a man-hour basis. From Table IX, Column 2, it may be seen that our measures show an increase of 290 percent in tires per man-hour between 1914 and 1927.

12 Bureau of Labor Statistics, Bulletin No. 616, *op. cit.,* pp. 707-736. For the period 1920 to 1929, the electric lamp industry showed a 240 percent increase in output per man-hour. In the tire industry the increase in productivity between 1921 and 1929 was 93 percent in pounds per man-hour and 410 percent in tire-miles per man-hour. It is probable that a similar allowance for the improvement in the quality of lamps would raise the rate of productivity in that industry above that of the tire industry.

13 Bowden states that between 1932 and 1936 output per man-hour increased 15 percent in all manufacturing industries and 16 percent in the tire industry (on a poundage basis). Ten of the sixteen industries studied by Bowden surpassed the tire industry in productivity increase in this period. These industries included steel, automobiles, cement, cotton goods, woolen and worsted goods, cigars and cigarettes, petroleum refining, crude petroleum producing, anthracite mining and class I steam railroads. Bowden, *op. cit.,* p. 22.

in the tire industry and in part with more general conditions affecting many industries in a more or less similar fashion. Both the conditions attending the increase in productivity and the incidence and effects of productivity changes may be expected to be more striking in the case of the tire industry than in other industries which have felt a smaller impact from the forces of technological change.

CHAPTER VI

FACTORS CONDITIONING THE INCREASE IN PRODUCTIVITY

General Factors

THE strong upward trend in productivity so greatly overshadows minor fluctuations of a seasonal and cyclical nature that the discussion in this chapter is devoted almost exclusively to an attempt to interpret and explain the trend of productivity. It has been suggested that the productivity of labor is merely a technical measure relating the quantity of goods produced to the amount of labor-time employed. The factors involved are so complex and so difficult to segregate that it is somewhat hazardous to discuss them as specific causes. Likewise it is quite impossible to determine the exact influence of each of them upon productivity. As a consequence, the factors concerned have been classified into groups and discussed in regard to their relative importance in terms of considerations attending the productivity increase. Although these factors are to a considerable extent similar in the tire industry and in other manufacturing industries, their relative importance is in large measure dependent upon the specific conditions marking the development of the tire industry.

The outstanding features of the tire industry are its relative newness and exceedingly rapid growth. Since 1910 the tire industry has grown from an unimportant subsidiary product of the relatively small rubber industry to a large mass production industry. It is only natural to expect a profound change in methods of production and in labor productivity in the transition from an output of less than 2,500,000 tires to an output of 55,000,000 or more tires per year. This same period has witnessed such revolutionary changes in the type and quality of tires that the effective increase in physical output in terms of tire-miles since 1910 has been more than 150-fold. Many of the changes in the processes of production which were instituted primarily in order to increase production and improve the product also resulted in increasing the productivity of labor. It is highly improbable that the physical volume of production could have been increased to anything like the extent that it has in the last quarter century without concomitant increases in productivity and reductions in labor cost.

The tremendous expansion of production and productivity has been made possible by a large inflow of capital into the industry. Although the emergence of the tire industry from its parent rubber industry obscures its early financial history, it appears that at the turn of the century carriage and bicycle tires were being manufactured in a large number of very small firms, the total assets of which probably did not exceed $1,000,000. Yet by 1929 the total assets of the tire industry are conservatively estimated to have exceeded $1,000,000,000. Although the next few years showed a decline, by 1938 the figure was approaching 1929 levels again.

A better measure of the role of capital in the industry may, however, be obtained by comparing the changes in the investment in fixed assets, including land, buildings, machinery and equipment. The net value of such equipment after depreciation is estimated to have been in excess of $250,000,000 in 1925, more than $340,000,000 in 1930, and between $250,000,000 and $300,000,000 in 1938. Another measure of the use of capital in the industry may be derived from data showing total capitalization of all firms in the industry, including bonded debt, preferred and common stock and surplus of each company. The Rubber Manufacturers Association estimated that the amount of such capital investment in the tire industry was $556,000,000 in 1929 and $419,000,000 in 1933. It is probable that the amount of such investment again exceeded $500,000,000 in 1938.[1]

1 Cross, Earseman, and Lenaerts, *op. cit.*, p. 18. The data were quoted from the N.R.A. Code Application for the tire industry made by the Rubber Manufacturers Association in 1933. The 1938 figures are rough estimates given orally to the writer by an official of the Association.

In an attempt to segregate the influence of capital investment upon productivity a study was made of the financial reports of a selected sample of the industry.[2] The movements of the ratios of fixed capital investment to production and employment between 1925 and 1935 were as follows:

Indexes: 1925 = 100

	1935	1932	1929	1925
Fixed Capital Assets [3]	115	122	138	100
Per Tire [4]	146	197	125	100
Per Pound [5]	108	165	120	100
Per Worker [6]	164	222	135	100
Per Man-Hour [7]	225	305	134	100

These data indicate that the ratios of fixed capital to production and employment advanced substantially but somewhat irregularly in the decade between 1925 and 1935. The conclusions which may be drawn from these movements are limited, however, by the fact that the industry had a large volume of unused capacity between 1929 and 1935.[8] The data seem to suggest that the growth of fixed capital contributed to the in-

[2] Financial data, compiled from the income tax returns of thirteen tire companies for the years 1925 to 1935, were supplied to the author by the National Research Project of the Works Progress Administration, acting in cooperation with the Bureau of Internal Revenue of the United States Department of the Treasury and the Bureau of Labor Statistics of the United States Department of Labor. The sample companies accounted for 75 percent of the rated capacity of the industry in 1933 and were selected to represent large, medium-sized, and small companies in proportions similar to the pattern of the industry in that year. The following analysis based on these data was made by the present author.

[3] Original cost less reserve for depreciation.

[4] Fixed capital assets divided by equivalent number of tire casings (Table I, Column 3).

[5] Fixed capital assets divided by crude rubber consumed (Table I, Column 4).

[6] Fixed capital assets divided by number of wage-earners (Table VIII, Column 2).

[7] Fixed capital assets divided by man-hours (Table VIII, Column 6).

[8] See Table VII.

crease in productivity, but that this development was chiefly a phenomenon of the period prior to 1930. The increase in productivity between 1929 and 1932 may be partly explained in terms of the increase in capital per unit of production and employment, but the large amount of unused capacity accounts for the greater part of the latter increases. The slackening of the rate of productivity increase since 1932 corresponds with the decline in ratios of capital to employment and production and it seems reasonable to assume that the former was in part caused by the latter development. In the period since 1930 it seems probable that such increase in productivity as was attributable to fixed capital investment was due largely to the displacement of the older less efficient units of equipment by the new and improved machinery purchased with new capital rather than to an increase in the total amounts of fixed capital used.

An indirect indication of the growth of capital in the industry over a longer period is afforded by the available horse-power data compiled by the Census of Manufactures, which shows 151,927 horse-power in the tire industry in 1914 and 463,257 units in 1929. Horse-power per worker nearly doubled between 1914 and 1929, the figures being 3.0 and 5.6 respectively.[9]

Another general factor which is probably responsible for an important part of the increase in labor productivity has been the concentration of the industry into larger and fewer establishments in the period between 1920 and 1933. In the first decade of the twentieth century tire manufacturing was just emerging as an industry, but by 1911 there were twenty-seven plants which were predominately tire producers. During the

[9] Data from E. G. Holt, United States Bureau of Foreign and Domestic Commerce, Circular Ru-3483, p. 2. The 1914 data are Holt's estimates of that part of the total horse-power of the rubber industry which was in the tire industry. Extreme caution must be exercised in interpreting horse-power data because of the shift from the use of steam to electric power during the period. The Census ceased collecting horse-power data after 1929 because of the difficulties encountered in their interpretation. It is very likely that horse-power data tend to underestimate the growth of physical capital in the industry.

next ten years tire demand increased enormously and by 1920 there were about 200 tire factories.[10] Between 1920 and 1935, however, the number of such plants was decreased by nearly 80 percent. In this period many small and inefficient plants were squeezed out while those that survived grew rapidly in size.

TABLE XIV

NUMBER OF ESTABLISHMENTS AND NUMBER OF WAGE-EARNERS PER ESTABLISHMENT IN THE TIRE INDUSTRY FOR SELECTED YEARS, 1911-1937 [a]

Year	Number of Establishments [b]	Wage-Earners per Establishment
1911	27	—
1918	111	—
1920	200	—
1921	178	312
1923	160	462
1925	126	648
1927	109	718
1929	91	915
1931	48	1024
1933	44	1204
1935	42	1360
1937	45	1334

[a] Sources: 1911-1920, estimated by Scudder, Stevens and Clark, *op. cit.*, p. 15; and Kress and Pearce, *op. cit.*, p. 7; 1921-35, Census of Manufactures; 1937 data were estimated by the writer, by a tabulation of the new plants built since 1935 and of the plants which ceased production since that time. See p. 169.

[b] An establishment is defined as an individual plant or factory.

Although many of the small concerns failed because of inability to withstand the heavy losses on inventories and the price wars, the higher labor productivity of the larger firms was an important factor in their survival and growth in this period. Moreover, in many cases where lack of adequate financial re-

[10] Some authorities estimate the number to have been considerably larger. The lack of agreement between estimates of the number of plants made by different authorities is largely due to dissimilar definitions. Many plants which were primarily engaged in the production of other rubber goods also produced some tires.

sources was the immediate cause of the failure of small concerns, the net result has been to increase the productivity of labor in the industry by shifting the business to the more efficient producers.

Mechanical Factors

It is, however, desirable to go behind the general factors conditioning the productivity increase in order to analyze the more specific reasons for the phenomenon. For this purpose, factors have been classified as " mechanical " and " non-mechanical ". The former set of elements refer to those which are directly associated with improvements in machines, whereas the latter include all other elements. From a theoretical point of view, the mechanical factors increasing productivity may be divided into two parts: (a) those involving an increase in the capital invested; and (b) those not involving an increase in capital investment. The first type is a change in the proportion of the factors of production or, more specifically, a substitution of capital for labor. The second type of change is a true improvement in technology reflecting the progress in invention and mechanical ingenuity. An examination of the history of the mechanization of the industry, or indeed of any given plant improvement, reveals that these two types of changes are almost inextricably intertwined. Consequently, mechanization is discussed as a single factor. As such, it is certainly among the more important forces involved in increasing productivity. The increased use of machinery and the innumerable changes in size, type and design of units have made tire production more rapid and automatic. This process has not only reduced the labor requirements in nearly all operations but it has made it possible to eliminate some operations entirely. The companies producing tire manufacturing equipment are continually designing new and improved units, and the engineers and research men of the tire companies are constantly studying the processes involved in an effort to effect improvements. There also has been a considerable amount of interchange of ideas and information among the research men and engineers of the several tire companies.[11]

Suggestions made by the workers themselves have also proved helpful. A workman who does the same job over and over again may visualize a change which will reduce the effort or time necessary for doing it. Most companies solicit workers' suggestions and pay bonuses of from $5 to $500 or more for those which are accepted. The chief difficulty is that several impracticable suggestions are received for every one which is used.

The mechanization of the tire industry, which has been responsible for so great a part of the increase in the productivity of labor, has not been the result of one or a few striking innovations in the process. It has been in the nature of a cumulative process resulting from a vast number of successive small changes, eliminating the labor of one man here and two or three there, so that a fairly long period must be observed in order to see the full effects of the movement. There has been no single fundamental change in the process of tire manufacture within the last decade, and in the last quarter of a century only one innovation can properly be considered revolutionary. This was the shift from the core process of tire building to the flat drum process, which was introduced in some plants as early as 1919 and was in use in nearly all tire factories by 1927. By the new process tires are built in the form of a flat circular band and later compressed into the shape of a tire in a vacuum box. This makes it possible to build passenger car tires in a fraction of the time required by the old core process in which the tire was built in its final shape.[12] Most tires for use on vehicles other than passenger cars are still built by the older method.

---

[11] The use of patented processes, of course, retards the tendency toward standardization of equipment and methods.

[12] Stern, *op. cit.*, pp. 7 and 24-25.

" Tire building " is the term used by the industry to describe the assembly of the various parts of the tire into the complete product. This operation is an important step in the process of tire manufacturing but should not be confused with it since tire manufacturing includes a great many other operations. See pp. 74-75.

FACTORS CONDITIONING PRODUCTIVITY 91

One other recent innovation, which may assume the character of a fundamental change, is the application of the assembly line technique to the process of tire building. In most plants the tire building operation is performed by skilled workmen and the entire process of assembling each tire is performed by one man. The work is done on tire building machines but the quality of the product and the speed of production depend upon the skill and dexterity of the workmen. The tire building machines, although power driven, are in reality merely complicated tools which facilitate the work. The new development in the process consists of setting a series of tire building machines on a platform which moves in front of the workers so that each workman may put one part onto a series of successive tires. It has increased the speed of the operation somewhat but the chief advantage of the new method lies in the possibility of substituting unskilled machine tenders for skilled workmen.[13] The United States Rubber Company in 1931 mounted a series of machines on a circular platform known as a " merry-go-round ". Although the " merry-go-round " is used exclusively by this company, the principle has been adopted in some of the newer plants of the other companies. The spread of the use of the process has been limited because at present it can be used to advantage only in plants manufacturing from 5,000 to 10,000 tires of the same size and type per day. It probably will be further developed and win more widespread adoption. If this happens it may prove to be a fundamental change for the industry.

In addition to the above mentioned innovations, a long list of less important specific changes in the process could be given. It is believed that the nature of the developments in the industry may be indicated to better advantage by classifying them into a few types of changes. Among the operations which have been

---

13 Stern, *op. cit.*, p. 52; and Drucker, *op. cit.*, pp. 57-58. It is stated that workmen can learn any of the jobs on the new basis in a week or ten days. In the old method it required about six weeks to learn the operations and from six months to a year for real proficiency.

most affected by mechanization are those involving the handling of materials within the factory. Elaborate systems of chutes, slides, and automatic conveyors have displaced the majority of the trucking operations formerly performed by hand. Moreover, the study of a plant made for the purpose of installing conveyor systems frequently leads to changes in the layout of the various departments which make the materials more readily accessible and improve the sequence of processes. Provision has also frequently been made for more convenient storage of materials and finished goods. These changes may result in considerable savings in the labor required for handling.

Another important set of factors is the increasingly automatic nature of the processes, as a result of which many relatively simple operations formerly performed by hand are now performed by machine by a slight alteration of the preceding or succeeding process.[14] It was in no small measure due to the accumulated effects of changes of this type that tire manufacturing has shown such a sharp increase in labor productivity. These changes have been both causes and results of many re-arrangements of the work within the factories. Several such changes have been combinations in sequence of two or more departments with a resulting reduction of the number of operators required.[15]

[14] Stern records a large number of such changes of which the following may be considered typical examples: New device for covering and flippering beads in one operation, 50 man-hours saved per day; Automatic stops installed on 4 stock cutting machines, 56 man-hours saved per day; Filling gum devices on band-building units, eliminating the need for rolling filling gum in liners, 240 man-hours saved per day; Bead flippering machine replacing hand processes, 240 man-hours saved per day; Cutting and rerolling departments consolidated and re-arranged, 112 man-hours saved per day. The list of such changes could be extended almost indefinitely. Stern, *op. cit.*, p. 53. The savings in man-hours relate to the plants studied by Stern. The application of each of these changes to other plants has saved labor, according to the scale of operations in each.

[15] The Ford tire plant at Dearborn, Michigan, which began operation in 1938, and is said to be the most efficient tire plant in the world, has made extensive use of the continuous flow principle. See E. V. Osberg, " The New Ford Tire Plant," *India Rubber World*, June 1, 1938, pp. 53-64, and advertise-

Another set of conditions making for an increase in labor productivity has been the use of larger units of equipment. Since in the majority of cases the larger units require no more operators than the smaller ones which they have displaced, the effect on productivity is obvious. In many instances the units for processing the materials, such as break-down mills, plasticators, mixing and rolling mills, and calenders are simply larger sizes of the same type. In other cases they are different types of equipment which process more materials in less time. The most outstanding example of this development has been the Banbury mixer, which has largely displaced the older roller mills in this operation. Clearly the larger plants have reaped an advantage from these changes at the expense of the smaller plants. Many of the financially weaker concerns also have not been able to introduce the newer equipment so rapidly. The concentration of a larger proportion of the business in a relatively few of the large companies has been both a cause and a result of these technological changes.

The acceleration of the chemical processes has also reduced labor requirements. Many such developments were results of the efforts to improve the quality of the tire, but they have had the incidental effect also of reducing the time and therefore the labor requirements in the processes involved. Accelerators were introduced largely to accomplish the latter end, however. Improvement in the quality of the raw materials, especially of crude rubber, has reduced the amount of labor required to prepare them for processing. The shift from the use of " pot heaters," which cured several tires at once, to the " watch case " vulcanizers, each of which cures one tire at a time, has been in the direction of smaller, more individualized units of equipment and contrary to the trend toward larger units. But this de-

ments of equipment used, pp. 65-68, and 50-52. Examples of combined operations are the Banbury mixers and sheeting mills, and calenders which coat both sides of the fabric with rubber in one operation. There are many other innovations, such as automatic weighing of most of the materials compounded with the rubber, feeding conveyors and the extensive use of automatic control boards.

velopment has contributed substantially toward increasing productivity in this department because the average time required for vulcanization has been nearly cut in half and because the handling labor in the department has been substantially reduced as a result of the change.

Non-Mechanical Factors

A. DEVELOPMENTS IN INDUSTRIAL MANAGEMENT

Many improvements in the management of production have contributed toward increasing labor productivity. The development of more precise materials specifications and the careful regulation of inventories of raw materials, goods in process, and finished goods have led to considerable reductions in labor required for handling. Studies of optimum size, layout, and engineering economy have led to substantial labor savings. Provision for regular, systematic equipment inspection tends to reduce the number of breakdowns. Production control, improved coordination of departments, and systematic routing of the goods in process, have also increased productivity by reducing the amount of labor's idle time. The disruption of schedules may be, to a large extent, prevented by the use of alternate routings and emergency crews. The Goodyear Tire and Rubber Company maintains a crew of especially well trained men, known as the " Flying Squadron ", whose function is to assist any department which is behind schedule. The effects of most of these developments upon productivity are known to be important in the aggregate even though they are not directly measurable. Two other developments in the field of industrial management, task standardization and the incentive wage, require more detailed treatment.

1. *Task Standardization*

Task Standardization has arisen from the principles of " scientific management " developed by Frederick W. Taylor in the closing years of the nineteenth century. His aim was to find " the one best method " of performing every task and to set it as a standard. This is now done by using the techniques of motion study and time study. Motion study involves a detailed analysis of every movement of the work of each job in a plant. The job is reduced to its basic elements and an effort is made to eliminate as much waste motion as possible by determining the easiest and most effective way of performing each specific necessary movement. Motion studies are not usually made until after the materials and equipment are standardized. Each motion study is followed by a time study in which the operators are timed by the steps or elements of the job. The standard time for the job is set at the average time of the average good operator, who has thoroughly learned the method developed by motion study. Daily tasks are determined by dividing the hours of work by the time per piece plus certain allowances for fatigue, delays, personal needs and other factors. After a plant has adopted task standardization, time studies are made periodically in an attempt to improve standards. Motion studies are usually made again only when a change is made in the amount or type of work done by a given workman.

Time and motion studies were made in a few plants in the tire industry before the World War but they did not find widespread application until the nineteen-twenties. The techniques are in use throughout the tire industry at the present time, with the exception of a few of the very small plants. Even the latter organizations achieve rough approximations of task standardization through the general experience of foremen and supervisors.

2. *Incentive Wage Systems*

Some sort of an incentive wage usually accompanies task standardization. The scope of incentive wages is somewhat larger than that of task standardization, however, since payment by result is practically universal for direct employees in the industry. The basic idea of the incentive wage is that the worker is paid for what he produces rather than for the time he is employed. Each worker earns more by producing more and thus he has a strong stimulus to increase his output. Since the specific devices for achieving this aim are legion, it is beyond the scope of the present inquiry to describe in detail the methods of incentive wage payment used in the tire industry. They may, however, be classified into three basic types, according to their effects upon the earnings of the individual worker. One type is the familiar straight piece rate, in which the worker's earnings increase in direct proportion to his output. A second type gives the worker a more than proportional increase in earnings for a given increase in output. The Taylor system is perhaps the best example of this type. A third type gives the worker an increase in earnings which is less than proportional to his increase in output. The Bedaux system, which is widely used in the tire industry, is of this type. Under this system it is usual to distribute 75 percent of the bonus for production above standard to the direct production workers and the remaining 25 percent to indirect labor and supervisory employees in the department.

All three of these types are subject to various modifications and many specific methods of wage payment are a combination of two or three of the types. Most plans have a standard performance requirement which is usually determined by task standardization and job analysis methods. In many cases the worker's earnings increase more rapidly at standard or above it than up to standard. In some cases the rates below standard are set in such a way that the poorest workers "fire themselves" but in most plans there is a "firing point". Some plans guarantee a minimum wage based upon time rates. Compensation usually is made to the workers at standard rates for time lost due to factors beyond their control.

3. Effects Upon Productivity

The effects of production control, task standardization and incentive wages upon productivity are practically inseparable since each is dependent upon the others. Their collective and cumulative effect upon productivity is recognized to have been very great. The task performances set for standardized shops

FACTORS CONDITIONING PRODUCTIVITY 97

are typically from 20 to 60 percent higher than previous averages. The shop averages usually exceed the first standards within a relatively short time, however, so that the productivity increase is usually greater than 50 percent. Charles W. Lytle states that under the Bedaux system the usual range of productivity increase after installation runs from 50 to 100 percent.[16] Many illustrations of productivity increases running as high as 300 percent may be found. Lytle concludes that the increases in productivity among skilled workers usually exceed those among unskilled workers when these methods are applied. He attributes this seemingly contradictory situation to the fact that prior to standardization unskilled workers are usually more carefully supervised than skilled workers.[17] When such large productivity advances are realized it may be necessary to reduce the force even though hours are cut and output is expanded in an effort to minimize unemployment. When employment is reduced usually the most efficient workers are retained and the selective process leads to further increases in productivity.

The incentive payment idea frequently is also applied to supervisory and managerial employees, thereby stimulating them to attain the most effective plant layout, control and routing. Changes resulting from such improvements and from task standardization may lead to further mechanization. In many instances an automatic device has been substituted for workers whose jobs have been reduced to relatively simple repetitive routines.[18]

B. SHORTENING OF THE WORK DAY AND WORK WEEK

Average hours of work per week declined from 49 in 1914 to 45 in 1923 and, after remaining nearly constant from that date until 1929, they then dropped rapidly to from 30 to 35 hours per

---

[16] *Wage Incentive Methods*, New York, The Ronald Press Company, 1929, pp. 250-255. These results relate to the Bedaux system as applied to any manufacturing industry. The system is used quite extensively in the tire industry.

[17] *Ibid.*, pp. 5-6.

[18] Stern, *op. cit.*, pp. 24-25.

week. Already by 1914 some of the leading companies were on the eight-hour day but many companies were still on ten- or twelve-hour schedules. It was about this time that continuous production on a 24-hour day basis was adopted in the industry, however, and three eight-hour shifts were so well adapted for this purpose that within the next few years most plants in the industry accepted the eight-hour day. In 1914 many workers in the smaller plants worked seven days per week but by 1923 the six-day week was well nigh universal and the Saturday half holiday was becoming increasingly common. By 1929 the eight-hour day and 44-hour week were standard for the overwhelming majority of tire workers, but longer hours still prevailed in some of the smaller shops.

Although statistical data are lacking, there is general agreement in the industry that the shortening of the work day and work week between 1914 and 1929 resulted in increased man-hour productivity which made the change well worth while for the industry as a whole, as well as for the workers. In 1932 Paul W. Litchfield, President of the Goodyear Tire and Rubber Company, wrote:

" Back in 1914 we went from the $10\frac{1}{2}$- and $12\frac{1}{2}$-hour day to the 8-hour day. At that time our study failed to find any increase in our overhead costs. It did, however, find a pronounced increase in efficiency, and our judgment is that the reduction from 8-hour days to 6-hour days would register a similar improvement if we were in position to push production ".[19]

In 1935 Mr. Litchfield made an even stronger statement when he wrote:

" In an eight-hour day a man could turn out more by the hour than he could in a ten-hour or a twelve-hour day, and with the aid of the machine, he could produce more in eight hours than in twelve—more and better products. *His efforts may have been the same—just as the effort in a hundred yard dash may be the same as in the quarter mile. But there was more money in the*

---

19 *Industrial Relations*, October, 1932. Quoted by Croxton, Lapp and Hanna, *op. cit.*, p. 2.

dash than there was in the run, more for the workman and more for the employer ".[20]

Between 1929 and 1933 average weekly hours declined from about 45 to about 32, largely because of falling production and the policy of " sharing the work ". The Goodyear Tire and Rubber Company instituted the six-hour day in its Akron plants in 1930, and within a relatively short time all of the other tire manufacturers in the Akron area, the Detroit plant of the United States Rubber Company, and several other tire plants throughout the nation likewise adopted it.[21] The President's Reemployment Agreement of June, 1933, and the N. R. A. code for the tire industry, which was in effect from December, 1933, until May, 1935, although not expressly providing for the six-hour day, greatly accelerated the movement in this direction.[22]

The cardinal advantage of the six-hour day is that the use of four instead of three shifts makes it possible for the industry to employ more workers. When the change was first made in most plants, the number of production workers was increased one-third. Since that time, however, both production and productivity have advanced so it is now practically impossible to determine the net increase in employment which is ascribable to the six-hour day. The relative advantages of the six- versus the eight-hour day from the point of view of efficiency of operation and labor productivity have given rise to considerable controversy in the industry.

[20] *Today*, February 19, 1935. Quoted by Croxton, Lapp and Hanna, *op. cit.*, p. 3. Italics supplied by the present author.

[21] Croxton, Lapp and Hanna, *op. cit.*, p. 1.

[22] *National Industrial Recovery Administration, Code of Fair Competition for the Rubber Tire Manufacturing Industry*, approved by the President, December 22, 1933. Under the code, hours of labor for factory employees were not permitted to average more than thirty-six per week for the year, nor in any case more than forty-two hours in any given week, nor more than eight hours in any given day. Overtime pay at the rate of time and one-third was specified for all hours over thirty-six in any given week. Higher limits on weekly hours were set for maintenance crews, watchmen, and salaried employees receiving less than thirty-five dollars per week. No limits were set on the hours of outside salesmen or salaried employees receiving more than thirty-five dollars weekly.

In the first instance most managerial officials took great pride in being able to minimize unemployment and at the same time increase man-hour productivity. A detailed study of the Bureau of Labor Statistics, made on the basis of data supplied by Mr. J. E. Lorentz, General Superintendent of the India Tire and Rubber Company, serves to confirm this. The experience of this small Akron company which made the change in the summer of 1931, first in the curing department and later throughout the plant, indicated that: (1) Labor costs per unit cured declined 8.2 percent; (2) Loss of production time decreased to less than 5 percent, although formerly it had been considerably in excess of this amount; and (3) Absences declined significantly.[23] Other managerial officials have reported hourly productivity increases of from 5 to 12 percent coincident with or as a result of the shift from the eight- to the six-hour day. The amount of time taken out for rest and lunch periods has been reduced and in several instances the organization of the operations has been changed in such a way as to reduce labor requirements. But, although it is recognized that increased man-hour productivity has accompanied the reduction in daily working hours, the phenomenon is subject to varying interpretations. The representatives of the unions state that the men worked faster as a result of the change and that the six-hour day is a cause of increasing productivity.[24] On the other hand many

[23] "Six Hour Shifts of the India Tire and Rubber Company," *Monthly Labor Review*, August, 1932, p. 369. These results relate mainly to the curing department but it was regarded as being reasonably representative of the plant as a whole.

[24] This idea was repeatedly expressed to the writer by union members and leaders. Many regard the "speed-up" as objectionable.

Croxton, Lapp and Hanna, discussing the controversy over the six-hour day at the Goodyear Tire and Rubber Company's Akron plants in 1935, stated:

"Representatives of A. F. of L. unions and the Labor Relations Committee of the Goodyear Industrial Assembly are unanimously of the opinion that employees generally "speeded-up" with the change from the eight-hour day to the six-hour day in 1930, and that in many operations where physical requirements are an important factor the older men in particular will not be able to stand the pace for the longer day. They state that the hardship on such employees is increased in many operations where there is a standard hour

managerial officials maintain that it is more correct to say that because the work day and work week were reduced it became necessary to raise productivity. Whichever view is adopted it seems probable that the productivity advance resulted partly from increased effort on the part of the wage-earners and partly from a variety of labor saving changes in the plants concerned.

The chief disadvantage of the six-hour day apparently is its inflexibility which makes it more difficult for the plants to adjust their operations and labor force to fluctuations in demand. Thus in the business spurt of 1936 and early 1937 the companies would have preferred to have lengthened hours to 40 or more before taking on more workers. Instead, in order to preserve the six-hour day, they were forced to hire more men whenever individual workmen reached 36 hours per week even when the plant averages remained below this figure Several hundred additional workers were taken on at this time, only to be laid off again later in 1937 and early in 1938. Although few of these workers were really new to the industry, many who had not worked in the industry for several months or even years had lost much of their former speed, skill and dexterity, so that labor costs were increased by employing them. The question of whether supervisory costs were actually increased by the six-hour day is very controversial. Croxton, Lapp and Hanna received much conflicting testimony on this point. Their conclusion on the matter was stated thus:[25] " The Fact Finding Board is of the opinion that supervisory costs need not be noticeably higher with a six-hour day than with an eight-hour day ".

Many managers who were in complete sympathy with work spreading arrangements in the acute stages of the depression now believe that the tire industry will not in the near future require as large a labor force as it did in 1929. These officials say

---

production, that is, where the output is in large measure controlled by the speed of the machine, by "pool" operations, or by some element of time." *Op. cit.*, p. 2.

See also Mr. Litchfield's statements quoted on p. 98 above.

25 *Op. cit.*, p. 2.

that the best policy for the industry is to provide full-time work at good wages for those to be retained and to dismiss the others, enabling them to seek jobs elsewhere. They point out that, although the tire industry may need a small labor reserve, it is not a healthy condition for it to have a large number of in and out workers. They also believe that weekly hours should be long enough to permit the workers to reap a considerable proportion of the benefits from hourly wage increases in the form of higher annual earnings. Several officials are of the opinion that a 40-hour week composed of five eight-hour days is the optimum work week if all factors—productivity, flexibility, supervision, and earnings—are considered.

The six-hour day was never accepted by a majority of the plants outside of the Akron and Detroit areas, and after the demise of the N. R. A., several plants which had once adopted it returned to the eight-hour day. In October, 1935, the Goodyear Tire and Rubber Company sought to do likewise in its Akron plant but the proposal aroused such widespread opposition among the workers that a serious strike was threatened. The newly formed United Rubber Workers union took a strong stand in favor of the six-hour day. It was supported by the International Association of Machinists, an American Federation of Labor affiliate which had some members involved. The unions petitioned the Secretary of Labor to appoint an impartial fact finding board as a preliminary step toward reaching an agreement. This board was appointed on November 15, 1935, and in its report of December 16, 1935, it recommended that the six-hour day be continued, chiefly on the grounds that the return to the eight-hour day would substantially lessen employment at a time when unemployment of tire workers was already widespread.[26] Goodyear accepted the board's recommendations.

[26] The Board members were Fred C. Croxton, John A. Lapp and Hugh S. Hanna. See *Findings and Recommendations, op. cit.* The controversy also concerned a proposed reduction in wage rates affecting about 27 percent of Goodyear's Akron employees. The board stated that the data at hand did not make it appear necessary to reduce wage rates. It recommended that wage

In March, 1938, the B. F. Goodrich Company opened negotiations with the United Rubber Workers in an effort to secure a more flexible work week. The company presented a proposal to establish the basic 40-hour week and eight-hour day but the terms were not accepted by the union and the six-hour day was retained.[27] At the present time all tire factories in Akron, the main tire plant of the United States Rubber Company in Detroit, and a few other tire plants throughout the country remain on the six-hour day and the basic 36-hour week. The majority of the tire plants outside of Akron and Detroit, including all the new branch plants of the Akron companies, are on the eight-hour day with a basic 40- or 44-hour week. In general the six-hour day is in effect only in those plants in which the United Rubber Workers union is strong enough to enforce this policy.[28] In these plants the usual policy is to reduce the hours of work to 24 per week before laying off any workers and to increase them to 36 hours per week before hiring additional workers.

C. THE SKILL AND EFFICIENCY OF THE WORKERS

Although changes in the morale, skill and efficiency of the workers themselves have not been the primary reasons for the increase in productivity in the industry, they are, none the less, important considerations. In the first place, the industry

adjustments be negotiated between the management and the representatives of the employees.

27 Statement of Mr. T. G. Graham in *The Akron Times-Press*, March 5, 1938. See also *The Akron Times-Press*, May 5, 1938. This newspaper and *The Akron Beacon-Journal* carried reports concerning the negotiations between the company and the union nearly every day between these two dates. The major question at issue concerned wage adjustments rather than hours, however. This controversy is discussed more fully on pp. 173-175.

28 In April, 1939, the president of the union stated: "Today the 6-hour day prevails in all tire plants under agreements with the union. The 8-hour day, 40-hour week is prevalent in other divisions of the industry."

Sherman H. Dalrymple, "The United Rubber Workers of America," *Labor Information Bulletin*, Volume VI, No. 4, United States Department of Labor, Bureau of Labor Statistics, Washington, Government Printing Office, p. 7.

has exercised more than usual care in the recruiting and training of its personnel and in the maintenance of a high employee morale. Tire workers as a group are better than average physical specimens largely because they have been selected partly on this basis. Although much of the heavy work has been eliminated or reduced by mechanization, many jobs in a tire plant still require unusual strength and stamina. This is particularly true of the "pit" or curing department, where it is still more unusual to find men weighing less than 175 pounds than to find them over 200 pounds in many factories. The industry has been able to secure superior workmen because it has long been among the highest wage industries in the country. This has also allowed tire workers to maintain a higher standard of living than is typical for industrial workers as a whole. The rise in the general level of education and the expansion of the welfare activities of the tire companies have also been contributing factors.

Nevertheless, the skill requirements for tire workers as a group have declined rather than increased in the period. Technological changes have operated to displace a larger proportion of the unskilled laborers than of the more skilled workmen; but mechanization, together with the application of mass production methods, have decreased the skill requirements for most of the remaining jobs. On the other hand, the skill requirements for some jobs remain substantially unchanged, while some of the more complicated machines require a few more highly skilled operators and more well trained repair and service men.

The industry does not make use of an apprenticeship system but rather trains the men on the job by special instruction and supervision. It is estimated that one-third or more of the jobs can be learned well enough to reach average efficiency within six weeks and that on at least one-half of them this stage can be reached within three months. From two-thirds to three-fourths of the jobs can be learned within six months and not more than five percent of them require more than a year to

FACTORS CONDITIONING PRODUCTIVITY 105

learn.[29] Within the last three years it has been demonstrated that the skill requirements for most jobs in a tire factory were less than generally realized. Several new tire factories have been started using only a very few experienced workers. Most of these plants have been able to operate at a fair approach toward normal efficiency within six months after their opening. At first their productivity was low, but within a relatively short time the productivity of these plants was approaching that of the plants using experienced tire workers.

Although on the average the skill requirements in the industry have declined, the tire workers have contributed substantially toward the increase in productivity. The workers' contribution has come about chiefly as a result of the speeding-up of the rate at which the average workman operates. There is no doubt but that the average worker in the industry at the present time has less idle time, fewer and shorter rest periods, and that his work is performed with less wasted effort and at a higher speed than twenty-five or even ten years ago. The mechanization of the process, the application of time and motion analysis techniques, and the wage incentive systems have been the contributing causes for the speeding-up of production, but the workers have put forth the additional effort required. It is not without significance that one of the most important issues raised by the United Rubber Workers is what they term the "speed-up." The increased speed of production has been achieved largely by the progressive raising of standard output

---

[29] Drucker, *op. cit.*, pp. 54-60. These percentages relate to the factory jobs only and, of course, do not apply to the considerable number of research men and technical experts employed by the tire companies.

There are substantial differences of opinion with regard to skill requirements, most of which hinge upon the definition of what constitutes learning the job. Admittedly many workers do not reach maximum efficiency in the above mentioned periods but it is believed that the new worker may be said to be trained when he begins to "make out," that is, to earn the basic hourly rate when working on piece rates. The more general adoption of assembly line methods of tire building may increase the proportion of unskilled workers considerably. See p. 91.

requirements on individual jobs as a result of motion and time analyses rather than by mechanical controls of output.[30]

The Influence of Labor Relations on Productivity [31]

Prior to 1933 no independent union ever succeeded in enrolling any considerable number of workers in the tire industry. Although many leaders of the industry insisted that there was no discrimination for or against union members, company pressure against unionism was undoubtedly a major factor. This attitude was more often implied than expressed but it was well understood by the workers. Probably more important as a deterrent to unionism was the establishment of personnel departments which attempted to maintain the working force at the highest possible levels of efficiency and contentment. In addition to the usual health and safety promotional activities, the personnel departments of several of the larger plants sponsored extensive social programs. Teams were organized and contests arranged in several sports, social rooms were maintained, dances, parties, and picnics were arranged and hobby clubs were developed. Clubs were organized also according to years of service with the company, as "ten-year clubs," "twenty-year clubs," etc.

Several of the companies were active in the provision of welfare services for their employees. Real estate developments, low-cost housing projects, credit unions, group life insurance, and pension plans were given financial aid. Study clubs and more formal courses of study were provided by some of the companies. The employee association, or company union, was sometimes adopted as a coordinating device for these various activities. Such organizations usually attempted to prevent the growth of unions more directly than an uncoordinated program of activities. Systematic procedures for settlement of individual grievances through shop committees were set up and the com-

---

30 Croxton, Lapp and Hanna, *op. cit.*, pp. 2-3; Stern, *op. cit.*, p. 25.

31 The main comparisons of productivity in this section are based upon pounds of crude rubber consumed per man-hour. See Table IX, Column 4.

FACTORS CONDITIONING PRODUCTIVITY 107

mon aims and interests of workers and management were stressed. One of the most successful of these organizations was the Goodyear Industrial Assembly, organized in 1919 and modeled after the structure of the United States government. It contained a House of Representatives and a Senate, elected by the workers, and dealt with grievances, wage adjustments and working conditions through a number of elective joint committees. Resolutions passed by the Assembly were sent to the factory manager for approval or veto. In the event of a veto, the measure, if repassed by the Assembly by a two-thirds vote, was submitted to the company's board of directors, whose decision was final. The plan contained a no-discrimination rule regarding union members but during most of the period of its active life very few union workers were employed by the company. The system apparently was reasonably effective in maintaining employer-employee relations and settling minor disputes for a period of about 15 years, but it did not prevent the growth of independent unionism in the period after 1933. All company financial support was withdrawn in 1937, however, following the Supreme Court decision upholding the National Labor Relations Act in its proscription of company unionism.[32]

Managerial opposition and company unionism were by no means the sole, and perhaps not even the chief, reasons for the lack of union organization in the industry prior to 1933, however. The rural or foreign origin of a large proportion of the workers, the anti-union tradition of Akron and of such secondary centers of the industry as Los Angeles and Detroit, the

[32] The scope of the present study does not permit a detailed analysis of the relative merits of employee associations and independent unions. It is obvious, however, that although the two types of organizations overlap in activities, they differ fundamentally in purposes and functions. The employee association is primarily a managerial device intended to keep the working force satisfied, whereas the independent union is primarily an association of workers for their own advancement. It is, therefore, almost inherent in the aims of the respective organizations that the former tends to stress the unity of labor and management in striving for a common goal, while the latter emphasizes the need for common action among the workers for the purpose of bargaining collectively with the management.

unfavorable experience of some workers with unions, and the fact that no unions were actively seeking members in the industry were also deterrent factors. The preoccupation of the American Federation of Labor with the craft form of organization, together with its notable lack of success in one or two early attempts at organizing the industry, explain the apathy of that organization toward the tire workers.

In 1902 and 1903 the first organizing attempt in the industry was made by the Allied Metal and Rubber Workers Union, affiliated with the A. F. of L. In the latter year the Amalgamated Rubber Workers Union of America split off from this group and was given a charter by the A. F. of L. The rubber industry at this time was still centered in the east and tires were an insignificant part of its products. About 300 workers were said to have been enrolled in Akron but the union died without even attempting much in the way of collective bargaining.[33] The American Federation of Labor staged another brief organizing campaign in the rubber industry in 1910 but the results were so meager that no records of its extent remain.[34]

In 1913 wage cuts in the major Akron plants gave rise to a group of apparently spontaneous strikes among unorganized workers. The Industrial Workers of the World, then near the height of their power, rushed organizers to Akron, began to enroll members, and took charge of the strike.[35] The A. F. of L. also sent organizers but succeeded in enlisting only a few members as the strike was already in other hands. The strike was conducted with a great deal of emotional fervor and a considerable amount of violence on both sides, but the employers were adamant in refusing all demands. After five or six

---

[33] Wolf and Wolf, *op. cit.*, pp. 498-499. The immediate cause of the demise of this union was the loss of a strike in a rubber plant in Trenton, N. J., in 1903. These authors state that the widespread use of labor spies was an important cause of the failure of this union to recruit any substantial number of members in Akron in this period.

[34] *Ibid.*

[35] *Ibid.* It is stated herein that the I.W.W. claimed to have enrolled 12,000 workers in Akron but that their actual enrollment was about 3,000.

weeks those workers who could get taken back resumed work at the employers' terms and union organization quickly disintegrated.

With these exceptions strikes appear to have been rare in the industry prior to 1933. By 1913 the industry had already entered upon its period of remarkable expansion. Demand increased so rapidly that for a time there was an actual labor shortage and the workers' demands in 1913, in so far as they related to wages and hours, were more than satisfied by 1920. Under these favorable conditions it is not surprising that labor relations were on the whole satisfactory both to management and to labor. There also seems to be no reason to doubt that the favorable state of labor relations was one of the many factors which made possible the large productivity increases realized in this period.

By the middle twenties, however, the expansion gave definite evidence of slowing down and the industry was distinctly less profitable than in the preceding decade. Employment was cut drastically in the 1921 depression and, although it was increasing again between 1922 and 1929, it never again reached 1919 proportions. The heavy mortality of small firms in the face of the competition with the larger producers increased the employment dislocations. At the same time that the larger firms were attempting to improve employee relations by welfare activities and employee associations, their scientific management programs were increasing the speed of operations.

Following 1929 the depression drop was unusually severe and all of these problems were magnified. Total unemployment was reduced by cutting hours but in many cases it was carried so far as to make partial unemployment very widespread. Even under these conditions the dissatisfaction of the workers was not frequently expressed in strikes. The Bureau of Labor Statistics reported only five strikes and 8,000 man-days idle, between 1927 and 1932.[36] Nevertheless, there is considerable

[36] Florence Peterson, *Strikes in the United State 1880-1936*, United States Department of Labor, Bureau of Labor Statistics, Bulletin No. 651, 1937,

evidence that unrest was increasing. Large numbers of workers had never received any benefits from company welfare and social activities, since these were undertaken extensively only by the larger and more prosperous concerns. No single company had ever engaged in all of the various activities heretofore described and, as the need for retrenchment became urgent, many such programs were curtailed or dropped. Mass unemployment was accompanied by wage cuts and constant pressure to increase productivity. The steps taken to enforce retrenchment often seemed arbitrary to the workers affected. Many plants had no machinery for the settlement of grievances except individual protest to the foreman, a procedure to which many workmen were reluctant to resort in times of such great job insecurity. Even in plants where employee representation systems were retained, the settlement of grievances became more difficult.

Thus it appears that the state of industrial relations was not as satisfactory in the decade of the twenties as in the preceding decade and that the difficulties were increasing between 1929 and 1932. The more or less paternalistic industrial relations proved satisfactory as long as the industry was highly prosperous and the workers were permitted to share the prosperity. The confidence imposed in management by large numbers of workers was weakened, however, by several years of hard times, which until 1933 appeared to be growing progressively worse. None the less, the state of labor relations remained favorable to increasing productivity throughout this period. Although the rate of increase in productivity was declining in the latter half of the twenties, it took a renewed spurt in the most drastic period of the depression, between 1929 and 1933.

The effort of each company to bolster its declining sales led to drastic price wars and increasingly heavy losses between

Washington, Government Printing Office, pp. 149-150. Only stoppages lasting two or more days were included. The reporting service of the Bureau was not complete during this period but it is believed that no large strikes went unrecorded.

1929 and 1932. In 1933 the National Industrial Recovery Act was adopted but the attempt to stabilize the industry under the Code proved largely unsuccessful. However, Section 7-A of the Act guaranteed the right to organize and bargain collectively and this seemed to offer the workers a means of bettering their condition. Following the promulgation of the President's Re-employment Agreement in 1933, the A. F. of L. lost little time in sending union organizers to Akron and other tire and rubber manufacturing towns. During 1933 and 1934, sixty-nine Federal Labor Unions with a total enrollment of about 30,000 workers were formed in the rubber industries.[37] The manufacturers responded by strengthening their employee representation plans while refusing to have any dealings with the outside unions. The union campaign thereupon lost steadily in membership in the latter part of 1934 and until about September 1935. By the latter date there were only 3,000 active union members in the industry.[38]

There seems to be no reason to doubt Dalrymple's statement that "The workers were organized in the face of the strongest opposition from employers."[39] It is likewise clear that the union drive largely disrupted the pattern of employer-employee relations which had prevailed in the industry since its beginnings and that discontent increased among the workers. In 1933 there were five strikes involving 18,000 man-days idle. Thus in this year alone there were as many strikes and more than twice as many man-days idle as in the preceding six years. In 1934 seven strikes resulted in 71,000 man-days idle, but in 1935 only one strike and 2,000 man-days lost were recorded.[40] Several other disputes which threatened to lead to strikes were settled by conciliation.

[37] Estelle M. Stewart, *Handbook of American Trade Unions*, 1936 Edition, United States Department of Labor, Bureau of Labor Statistics, Bulletin No. 618, Washington, Government Printing Office, pp. 223-224.
[38] Dalrymple, *op. cit.*, p. 4.
[39] *Ibid.*
[40] Peterson, *Strikes in the United States, 1880-1936, op. cit.*, pp. 149-150.

In spite of the somewhat disturbed state of industrial relations in this period, the Federal Labor Unions achieved relatively little in the way of collective bargaining. The direct concessions which they gained in matters of wages and hours were relatively minor. Some of the strikes were intended to secure the recognition of the unions as bargaining agents but here also their gains were small. The chief accomplishments of the unions in the tire industry in this period came as a result of their participation in the N. R. A. code formulation and administration. The decline in union enrollment in the industry was under way before the N. R. A. was declared unconstitutional in the spring of 1935 but it was greatly accelerated in the four months following that event.

Ever since the beginning of the union drive in the industry in 1933, the leaders of the rubber workers had been pressing the American Federation of Labor for an international charter on an industry-wide basis. In September, 1935 the Federation yielded to their request and the Federal Labor Unions of rubber workers joined together to form the United Rubber Workers of America. This organization was given jurisdiction over all workers in the rubber industries. At the time of the union's formation the first national convention was held in Akron and plans were laid for a new organizing campaign.[41] But the new union was not large enough to finance the drive on the scale which it deemed necessary, so it turned to the Federation for assistance. Meanwhile, the officers of the Federation watched the mounting conflict between its industrial and craft union affiliates with many misgivings. The former group was pressing the Federation for financial support for an industrial union drive in the rubber, steel, automobile, and other mass production industries, while the latter steadfastly opposed such action.

Gradually in the closing months of 1935 it began to appear that the group in opposition to the drive was winning. Mean-

---

41 *Proceedings of the First Constitutional Convention of the United Rubber Workers of America, Affiliated with the American Federation of Labor,* Akron, Ohio, September 12-17, 1935.

while in November, 1935, the heads of a few unions which were organized on an industrial basis joined together informally as a Committee for Industrial Organization under the leadership of John L. Lewis. In February, 1936, when the United Rubber Workers conducted a strike against the Goodyear Tire and Rubber Company in Akron, they received considerable assistance from the Committee for Industrial Organization. The United Rubber Workers joined the C. I. O. at this time. It is generally regarded as one of the ten pioneer C. I. O. unions, as it was among the nine unions which were first suspended and then expelled from the A. F. of L. after peace efforts between the rival federations failed. The United Rubber Workers also participated in the reorganization of the C. I. O. as the Congress of Industrial Organizations in 1938.[42]

In 1936 no less than 22 strikes were called and 444,000 man-days were lost in the tire industry. The disturbed period lasted into 1937, in which year there were ten strikes involving 526,000 idle man-days.[43] This was the period of the great organization drive of the United Rubber Workers under C. I. O. auspices. The new organization demanded the right to organize and bargain collectively which had been guaranteed to labor by the National Labor Relations Act of 1935. It demanded recognition as the sole bargaining agent wherever its membership included a majority of the employees of a given plant. Although this was in accordance with the administrative decisions of the Board regarding "majority rule", it was a matter of bitter contention in several strikes in the tire industry. The union also pressed its other demands vigorously. In addition to wage concessions it fought for recognition by a written union contract, the elimination of all discrimination against its members, the complete abolition of company unions, the six-

---

[42] Dalrymple, *op. cit.*, p. 4.

[43] Peterson, *Strikes in the United States, 1880-1936, op. cit.*, pp. 149-150; *Analysis of Strikes in 1937*, United States Department of Labor, Bureau of Labor Statistics, Serial No. R. 789, 1938, Washington, Government Printing Office, p. 6. The 1937 strikes were concentrated largely in the first six months of the year.

hour day, and seniority rules. The "speed-up" was vigorously denounced as the union sought a voice in the determination of standards of individual output. In some cases it also sought the "union shop."[44]

The tire and rubber workers have frequently been given the somewhat dubious distinction of introducing the sit-down strike into the United States.[45] The widespread use of this paralyzing technique contributed much to the success of the new union, which won nearly all of its major strikes in 1936 and 1937. In achieving these victories the union was also assisted by a combination of other favorable circumstances, including the protection of the right to organize and bargain collectively afforded by the new federal legislation, a strong upsurge in business activity, vigorous union leadership, and a well financed organizing campaign. As a result the membership of the United Rubber Workers grew rapidly. By the end of 1935 the total union membership was not more than 5,000 but by September, 1936, 25,000 members were enrolled, and in September, 1937, the union laid claim to 75,000 members. The

[44] The "union shop" permits the employer to hire whom he pleases but requires all new employees to become members of the union within a specified time. It is somewhat less rigid that the "closed shop" which requires the employer to hire union members only.

[45] Wolf and Wolf, *op. cit.*, p. 517. These authors consider the 58-hour strike at the Akron plants of the Firestone Tire and Rubber Company in 1935 the first major sit-down strike in this country. Actually the technique is not a new one, since stoppages in which discontented workers remained idle in their work places have been known for at least a century in both Europe and America. Such disputes usually lasted only a few hours, however, and after that time the workers either returned to work or left their work places to conduct the conventional type of strike. Several such stoppages occurred in the tire industry in 1934 and 1935. The new element introduced by the sit-down strike appears to have been the maintenance of possession of the factories by strikers for a longer period. Although the technique preceded the C.I.O., the United Rubber Workers under the C.I.O. affiliation made use of it on a much larger scale in 1936 and 1937, and the sit-downs soon spread to other industries. Such strikes were widely condemned as unlawful violations of property rights, however, and the device has not been used in any strike of importance in the tire industry since the latter part of 1937. Its use has also died out elsewhere in the United States since about that time.

FACTORS CONDITIONING PRODUCTIVITY   115

number of locals increased from fewer than 20 in 1935 to 132 in 1938.[46] By no means all of these persons are employed in the tire industry, however, because the union seeks members among all workers in the rubber industries. Approximately half of the wage-earners in the rubber industries are employed in the tire industry, but since a larger proportion of the tire workers are organized they constitute a majority of the union. In spite of the large number of members and locals, almost half of the membership is contained in the plants of the six tire companies in Akron and vicinity and in the main tire plant of the United States Rubber Company in Detroit. The union at first concentrated upon organizing the more highly skilled tire workers, who formed the backbone of the organization, but the membership now includes a much larger number of unskilled or semi-skilled than of skilled workers.

Although it is difficult to appraise the effect of industrial relations upon productivity because of the number and complexity of the factors involved, it none the less seems reasonably clear that the state of labor relations since 1933 has been less favorable to advancing productivity than prior to that time. The failure of productivity to advance in 1933-1934 coincided with the American Federation of Labor organizing campaign, which disrupted the old pattern of industrial relations. No doubt the unsettling effects of the union drive were in part responsible for the lack of productivity increase, but it must also be recalled that most of the factors which contributed so largely to the upward trend of productivity were also absent. The low ebb of production in this period, the absence of plant modernization or improvement, and the adjustments which the industry

---

[46] *Proceedings of the First Convention of the United Rubber Workers of America, Affiliated with the Committee for Industrial Organization*, Akron, Ohio, September 13-21, 1936, and *Proceedings of the Second Convention of the United Rubber Workers of America, Affiliated with the Committee for Industrial Organization*, Akron, Ohio, September 12-20, 1937.

Dalrymple states that a peak of nearly 80,000 was reached in 1937, but that the decline of employment by nearly 40 percent in 1938 brought with it a decrease in union membership of about 15 percent, *op. cit.*, p. 4.

was making to the N. R. A. code were strong deterrents to productivity.

Similarly the resumption of the upward trend in productivity in 1935 coincided with a decline in labor disputes. Here again these factors were probably associated but other factors, such as the return of a more hopeful business outlook, the resumption of technological improvements, and the release from the restrictions of the N. R. A. also contributed toward increasing productivity.

The renewed organization drive in 1936 and 1937 was marked by more serious disturbances in the industry. The failure of productivity to advance in 1936 in spite of a considerable increase in production and generally improved business conditions was probably in large measure attributable to the disturbed state of industrial relations. The industry remained torn by the major industrial conflicts in the early months of 1937 but since that time labor relations have greatly improved. In 1937 productivity increased once more in spite of the unfavorable labor situation in the first few months and the business downturn in the latter part of the year. It seems probable that this decline in production and employment, which lasted throughout most of 1938, tended to reduce labor conflicts because neither the union nor the companies could afford to precipitate a major struggle. The upturn of 1939 has not led to an increase in labor disturbances, however.

By 1938, the union either had an exclusive bargaining agreement or more or less satisfactory working arrangements for collective bargaining with most of the major plants and several smaller ones in the industry. The unions and the managements in these plants were beginning to work out their problems in an atmosphere, which, although not intimate, was much improved. The unions continued to press their demands for written collective bargaining agreements, higher wages, the six-hour day, and the adjustment of individual grievances. The part of the union program, however, which has most directly affected productivity has been the campaign against the " speed-

FACTORS CONDITIONING PRODUCTIVITY 117

up." It has been the contention of union leaders that the use of motion-time analyses, wage incentive systems, and other devices have increased the pace of the average tire worker to such an extent that it is damaging to the workers' health and strength. As a consequence, various local unions have adopted rules opposing the further speeding-up of production. These restrictions are the source of some of the bitterest differences of opinion between managers and union leaders, as well as among the primary objections of many industrialists to the whole union movement. It appears that the national union has no such rules but that many of the locals make and enforce them. Union leaders admit that they have been unduly restrictive in some instances, but these cases are regarded as mere temporary manifestations of misplaced zeal on the part of workers new to the labor movement.

The effects of restrictions upon production have shown up sharply in some few departments and plants, and they have on the whole undoubtedly reduced productivity somewhat. They appear to be on the decline, however, and they probably do not now seriously retard the upward trend of productivity. The rise of a vigorous union movement has undermined the confidence between management and labor, but as the labor movement becomes more seasoned this confidence is slowly being rebuilt on a new basis. Likewise, although the coming of the union and new industrial conflicts tended to undermine the pride in skilled craftsmanship and substitute for it the idea of labor solidarity, the union is now stressing pride in workmanship although retaining the spirit of solidarity. One result of unionism which remains, however, is that the present day workmen are much more " dollar minded " than before and will press for every advantage calculated to increase their earnings.

As will be seen in a later chapter, the indications are strong that the growth of unionism has strengthened the tendency toward the decentralization of the industry. Moreover, the emergence of labor relations as a troublesome managerial problem and the rise in wages and labor costs probably have stimulated some managements to economize on labor by substituting automatic processes whenever feasible. Such a trend appears to be in evidence in the new Ford tire plant and in the new branch plants of the major producers, but there is not much evidence of these developments in the older, established plants in the industry. This development would undoubtedly have proceeded much further than it has to date if management were not seeking so industriously to rebuild labor good will. One of the best means for gaining labor's confidence is to use labor instead of machinery wherever the cost advantages are nearly equal.

CYCLICAL AND SEASONAL INFLUENCES ON PRODUCTIVITY

The strong upward trend of productivity makes it difficult to detect cyclical manifestations. Morever, a thorough analysis of cyclical influences on productivity would require monthly data for an extended period. Since man-hours on a monthly basis are not available for a period long enough to make a significant analysis it is necessary to rely upon the annual data. Since 1923 the trend of productivity in terms of pounds of crude rubber consumed per man-hour was as follows: [47]

Years	Percent of Increase	
	Total	Average per Year [48]
1923–1929	53	7.6
1929–1933	72	14.4
1933–1937	17	3.4

During the period from 1923 to 1937, productivity in pounds of crude rubber consumed per man-hour advanced in 12 years, declined slightly in two years (1926 and 1934), and remained stationary in one year (1927). In 1936 the productivity increase was insignificant. The net increase over the period was 211 percent, but the rate of movement varied considerably. Productivity apparently advanced more rapidly in

[47] These data were derived from Table IX, Column 4.

[48] Total divided by the number of years. The average percent of increase per year is based on the beginning year in each case and is *not* cumulative.

the recession and depression years between 1929 and 1933 than in the prosperity years 1923-1929. In the recovery years 1933-1937, when business was not very good but was improving, the productivity increase was much less than in either of the other two periods. The recent retardation in the rate of productivity advance was not, however, unprecedented. The 1923 to 1929 period was not entirely homogeneous from the point of view of either production or productivity. The increase in man-hour productivity was 37 percent (12.3 percent per year) from 1923 to 1925, but only 12 percent (2.4 percent per year) from 1925 to 1929. Production in pounds of crude rubber consumed increased 51 percent from 1923 to 1925, but the increase amounted to only 15 percent from 1925 to 1929. The amount of crude rubber consumed declined 8 percent between 1925 and 1926 and an additional one percent in 1927, but rebounded sharply in 1928 and 1929. On the other hand, production declined 29 percent between 1929 and 1933, while productivity increased rapidly. Likewise, the amount of crude rubber consumed increased 47 percent between 1933 and 1937, but productivity increased much more slowly than in the preceding five years. Thus productivity increased rapidly with a rapid increase in production from 1923 to 1925, but it increased at a slightly greater rate when production was falling rapidly from 1929 to 1933. On the other hand, productivity increased slowly while production was see-sawing upward at a moderate rate from 1925 to 1929, and it also increased slowly while production was advancing rapidly from 1933 to 1937. Perhaps monthly data would show a more regular pattern if they were available, but the annual data are too inconclusive to establish any clear relationship between cyclical movements of production and productivity.

The different stages of the business cycle do, however, bring with them conditions which alter the relative importance of the factors conditioning the increase in productivity. During recovery and prosperity periods several factors operate to increase productivity. New capital flows into the industry and new machinery and equipment are installed most rapidly in the upward movement of the cycle. Although the business cycle does not appear to affect greatly the actual invention or discovery of new techniques, undoubtedly more money is spent on research during prosperity. Moreover, new techniques which are developed in prosperous times are usually adopted almost immediately, whereas those which are developed in depression periods do not usually find wide application until recovery is under way.

Productivity advances in recession and depression periods also partly as a result of the same factors which operated in recovery and prosperity and partly as a result of different ones. Plant modernization and mechanization are halted as the volume of incoming capital is reduced to a trickle, but the fruits of much of the investment and research of the late prosperity period mature more fully in the recession and depression phases of the cycle. There is, moreover, a rigidly selective process in which the output of the least efficient plants is reduced most. Nearly every depression forces some high cost producers out of the market, at least temporarily, and the severe depressions usually force some plants out of business permanently. Likewise, it has been usual to retain only the most efficient workers during these times.[49] In addition the shorter hours worked tend to stimulate workmen to greater efforts because each wishes to

---

[49] The growing importance of seniority rules, under which the order of lay-offs and rehirings is determined by length of service, now limits the operation of this selective factor. The United Rubber Workers insist upon the incorporation of such a provision in their trade agreements. Several other plants in the industry have similar rules in effect. The union favors the policy because it protects their members from arbitrary discharge and gives them a greater degree of job security as they grow older. Many managers believe that the disadvantage of not being able to follow the selective policy is more than outweighed by the improvement in employee morale resulting from the increased security. As a worker's length of service with a company increases, his stake in his job and in the welfare of the company upon which it depends, is increased. On the other hand, some managers insist upon the right to hire, fire, and lay-off men on merit alone, and some of the younger workers object to seniority rules because they place the brunt of the unemployment upon the newcomers.

earn as much as possible during the limited number of hours he has to work. The piece rate method of wage payment strengthens this tendency. This stimulus to greater production has been limited to some extent by the union restrictions against the "speed-up." Likewise, management is stimulated to simplify processes, to drop those least necessary, and to bend every effort toward reducing costs during periods of declining production. The pressure to minimize losses sometimes evokes greater responses than the drive to increase profits.

The tire industry has a fairly pronounced but rather irregular seasonal swing, with production usually reaching a peak in May or June and declining to a low in November or December. Over the past fifteen years production in the peak month has averaged 40 percent above that of the lowest month, but the variation in number of workers employed has been kept to between 10 and 15 percent in most years by means of varying the weekly hours. Since the productivity data used in this study were on an annual basis the effects of seasonal variations have been eliminated. Supplementary test computations fail to show any significant productivity variations from the peak to the dull seasons. It is known from general observation, however, that installations of new equipment and major changes which disrupt the flow of production are made in the slack season in so far as possible. Thus it appears that the industry has adjusted its schedules to the seasonal fluctuations so satisfactorily that the variations in productivity which may be ascribed to this factor are negligible.

CHAPTER VII
SOME EFFECTS OF INCREASING PRODUCTIVITY

Effects Upon Employment

PERHAPS the most striking effects of the tendencies described thus far have been upon the employment afforded by the industry. Consideration of these effects may be facilitated by dividing the history of the tire industry into three parts: (1) From its beginning to 1919; (2) From 1920 to 1929; and (3) From 1930 to the present. These dates not only mark off rather well defined periods of the history of the industry but they have the added advantage of corresponding to distinct periods of American economic history.

The first period was one of rapid growth in both production and employment. Since tires were a new product the output increased from literally nothing shortly before the turn of the century to 32,000,000 tires (154,000,000,000 tire-miles) in 1919. Thus, within a generation there arose an industry, which in 1919 gave employment to nearly 120,000 persons, of whom more than 89,000 were employed in making tires. In this period there can be no doubt but that the increase in the productivity of labor greatly increased employment. For just as the pneumatic tire is one of the fundamental inventions upon which the automobile industry rests, so is it highly improbable that the automobile could ever have reached its present development without the great strides which have been made in producing better and cheaper tires.

The tire industry enjoyed a period of prosperity in the pre-war years and in the war boom, but in the closing months of 1920 the collapse of the war-time prosperity gave it a hard jolt. The number of wage-earners employed in manufacturing tires dropped from 89,000 in 1919 to less than 51,000 in 1921.[1]

[1] The Census of Manufactures reclassified the industry in 1921, so that the above figures are more indicative of the real decline in employment than

EFFECTS OF INCREASING PRODUCTIVITY   123

The industry revived again in 1922, however, and employment continued to rise until 1929. Nevertheless, in spite of the fact that more than twice as many tires and nearly 7 times as many tire-miles were produced in 1929 as in 1919, there were 17,000 fewer employees manufacturing tires in 1929 than ten years earlier.

Between 1929 and 1932 the number of wage-earners manufacturing tires declined by no less than 32,000. Between 1932 and 1937 approximately 10,000 workers were re-employed, but the 1937 level remained 22,000 below 1929 and 39,000 below 1919. This represents an annual net displacement of 2,167 workers over the past 18 years. Although the lay-offs were heavily concentrated in the years 1920 to 1922 and 1929 to 1932, the subsequent revivals from these depressions did not restore employment to previous levels.[2] The decline in employment stands in sharp contrast to the production record of the industry, which in 1937 reached a new peak in terms of rubber consumed and tire-miles produced (although not in terms of number of tires produced). The indications are that the tire industry will not in the near future reach the 1929 level of employment, to say nothing of the level of 1919.

The situation of the tire industry corresponds to that of many manufacturing industries. It has led some observers to take the view that our economy is now confronted with a new phenomenon known as technological unemployment. This term may be taken to mean the loss of employment due to a major decline in the demand for labor, especially when resulting from changes in the technical methods of production which reduce unit labor requirements. The increase in the productivity of labor is held by many persons to be one of the major causes of unemployment and the experience of the tire industry is held to be a typical case in point.

the Census figures for the industry which give 120,000 employed in 1919 and 55,000 in 1921. See Chapter IV.

[2] The figures quoted above relate to employment in the manufacture of tires and tubes only. The tire industry employment figures include persons employed in the manufacture of various other products made in tire factories.

Technological unemployment, however, far from being new is merely a new name for a phenomenon at least as old as the industrial revolution itself. Moreover, history reveals that in the long run the increase in the productivity of labor by mechanical or other means, has vastly increased the opportunities for employment. The tire industry affords an example of this development. The invention of the tire, the improvements in its quality, and the increase in the productivity of labor in tire manufacturing are responsible for the fact that in 1937 some 50,000 workers were employed in making tires. This is approximately 50,000 more than were so employed in 1890, although it is 39,000 less than in 1919. Whether technological unemployment exists depends upon the years for which the comparisons are made.

An increase in the productivity of labor makes it possible for an economic society to produce more goods and services and to raise the standard of living. These changes, although manifestly in the general interest, do undoubtedly work serious hardships upon the individuals who lose their jobs as a result of them. The number of workmen who have lost their jobs in the tire industry as a result of technological changes increasing the productivity of labor has been large. It has in fact been considerably in excess of 39,000 in the last 18 years because that figure relates to the net change in employment only. These workers are forced to seek other jobs either in the tire industry or in other industries. Their success depends upon business conditions both in the tire industry and in the other industries of the nation.

Prior to 1919, although productivity was increasing rapidly and large numbers of workers lost their jobs as a result of technological changes, the industry was increasing its labor force, and most of the workers were reabsorbed in the industry. Some workers experienced the hardships of temporary unemployment but increasing productivity expanded employment in the industry as a whole. Between 1919 and 1929 productivity in the tire industry increased more rapidly than production, so that there was a net decline in employment in the industry. At the same time, however, employment was expanding in many other industries and presumably most of the displaced workers succeeded in finding other jobs. Since 1929 employment in the tire industry has declined further, but the workers displaced have had much greater difficulty in finding other jobs because of the generally poor state of business. In this period the suffering of the technologically displaced workers has been most severe. The difficulty appears to lie, not with the tire industry, but with the conditions which have made employment opportunities so scarce throughout the economy.

The remedy is not to be found in forcing the tire industry (or other similarly situated industries) to use their labor forces less efficiently in order to create more jobs. To attempt this would be to strike at the very roots of progress. Likewise, it appears that not much more can be done in the direction of spreading the work by shortening the weekly hours of labor in the tire industry. One constructive step is to encourage the development of new industries to absorb the unemployed. Another is to correct, in so far as possible, the conditions which make our economy subject to withering depressions. Further discussion of these programs is clearly beyond the scope of the present work. A more pertinent phase of the problem for present purposes, however, results from the fact that the adjustments of the economic system, which affect employment so vitally, are closely related to the distribution of the gains resulting from the increase in productivity. Both are intimately tied up with the workings of the price mechanism which serves as the adjustor of the various activities of our complex economic system.

The Distribution of the Gains of Productivity Increase

When a technological advance is made which reduces the labor required to produce a tire, there is a net gain to some group or groups in the economy unless it is offset by an increase in other elements of cost. With the data available it has been found impossible to determine exactly the extent to which the gains from increasing productivity have been counterbalanced by such other factors. However, some light may be thrown on the problem by an analysis of certain expense items of ten tire companies between 1925 and 1935. In 1935 the industry's production was 18 percent below that of 1925 in terms of equivalent number of tires, but 6 percent above 1925 in terms of rubber consumed, and 57 percent above 1925 in terms of tire-miles produced. If it may be assumed, as seems probable, that the production of these ten companies varied in about the same manner as that of the industry as a whole, these measures of production may be compared with the following expense items of these companies, each of which is given for 1935 as a percentage of the 1925 figure.[3]

	1935	1925
Administrative and Selling Expense	178	100
Capital Expense	97	100
Wages and Salaries (Manufacturing Only)	74	100
Materials Expense	33	100

Thus it appears that administrative and selling expenses increased between 1925 and 1935 at a more rapid rate than production, even if the latter be measured in tire-miles. The more detailed data indicate, however, that the increase in these items was to a large extent attributable to the development of manufacturer-owned retail outlets. To the extent that this represents an absorption by the manufacturer of functions formerly performed by independent retailers this is not a real increase in the cost of tires to the consumer. Manufacturing wages and salaries

[3] Based upon an analysis of the income and expense statements of ten tire companies in 1925 and 1935. These companies were part of the sample study described on p. 86. Three companies of the original sample could not be included in this analysis because their statements were not given in sufficient detail. The ten companies represent 57 percent of the capacity of the industry as of 1933. The data relate to total expenses of the companies studied in these categories not to unit costs. Capital expense includes depreciation, repairs, maintenance and tool expense. The item thus relates to the expense incident to the upkeep of physical capital and is independent of the methods used in financing the companies.

fell off less than might have been expected in view of the 30 percent decrease in employment and nearly 49 percent decline in man-hours during the interval. Capital expense declined slightly during the decade but since production declined to a greater extent it appears that unit capital costs increased somewhat. But the most significant element of all is the decrease in material costs by two-thirds. Material costs exceeded the total of all elements of cost combined in 1925 and they were still much the largest single element of costs in 1935. The decline in the cost of materials more than offset any tendency for other costs to rise, so that the trend of costs other than labor costs has been downward since 1925.

The increases in productivity have resulted in reductions in costs and have made it possible for certain groups within the economy to receive net gains. Likewise, the net reduction in elements of cost other than labor have made additional net gains available. Since it is impossible to separate the gains from productivity from those derived from other sources, the following analysis is based upon the distribution of the gains from net cost reductions regardless of source. These gains may be distributed to the consumers, to the wage-earners, or to the owners and managers of the industry. If the gains accrue to consumers in the form of lower prices, more tires may be sold and the expansion of production may reabsorb the displaced workers. This is the chief way in which increasing productivity has contributed toward the growth of the industry. If, however, consumers in spite of the lower prices do not expand their purchases proportionately there is a net release in purchasing power which will lead to the expansion of other industries. The increase in employment in these industries may offset the decrease in employment in the tire industry. Likewise, if the gains from productivity are passed on to other groups they may expand their purchases either of tires or of other commodities. In this way the workers displaced by technological changes are normally expected to be reabsorbed into industry. These adjustments work themselves out more or less satisfactorily through the price system. Prices must be flexible, however, and the factors of production must move freely under the stimulus of competition in order to have the system function properly.

Professor Frederick C. Mills has summarized the conditions necessary for the smooth functioning of such adjustments as follows:

"The working of a modern economy involves contact between productive energies and purchasing power. A gain in productivity brings, on the one hand, a release of productive energies; less effort is required for the production of a given quantity of goods. On the other hand, a gain in productivity brings (almost invariably) a shift in the distribution of purchasing power, a shift from group to group or a change in the distribution within a group. The essential problem of readjustment is that of establishing a connection between the released productive energy and the purchasing power thus shifted. Under contemporary conditions this connection can be established only through the system of prices. Whether it be directly and readily established, with only temporary dislocation, or whether the process be long drawn out, with protracted unemployment of productive resources, depends in good part upon the original incidence of the change and upon the number and the character of the frictions retarding a realignment of productive agencies and a readjustment of producer-consumer relations. *In an economy marked by numerous frictions the pains of readjustment may be minimized, as I have suggested elsewhere,[4] when the gains of industrial progress are widely diffused.* When the gains are narrowly restricted the economy may operate for an extended period at an inefficient level."[5]

[4] See *Prices in Recession and Recovery*, National Bureau of Economic Research, New York, 1936.

[5] "Industrial Productivity and Prices," *Journal of the American Statistical Association*, June, 1937, Volume XXXII, pp. 247-262. The quotation is from p. 262. The italics were supplied by the present author.

A. LABOR COSTS

The analysis of the gains to various groups resulting from the productivity increase is made difficult by the lack of adequate data and the intricate interrelationships between labor costs and other cost elements. Labor costs constitute only from 13 to 18 percent of the total costs of production of a tire.[6] An increase in productivity which is not offset by increases in other costs may lead to either an increase in wages or a reduction in labor costs. Thus the gains from increasing productivity which accrue to all groups except the wage earners must come from the reduction in labor costs. The reduction in labor costs since 1914 is shown in Table XV. Although these figures are merely estimates in which the margin of error may be quite high, the trend is unmistakable. Labor costs per tire have been reduced nearly 30 percent and labor costs per tire-mile have been reduced nearly 88 percent since 1914.

TABLE XV

Labor Costs per Tire and per Tire-Mile for Selected Years, 1914–1937

Indexes: 1914 = 100

	Labor Cost	
Year	Per Tire [a]	Per Tire-Mile [b]
1914	100.0	100.0
1919	158.9	118.2
1921	113.5	72.6
1925	76.6	27.0
1929	73.3	17.2
1932	47.4	9.4
1935	64.4	11.6
1937	70.6	12.4

[a] Tire industry payroll index from Census of Manufactures and Bureau of Labor Statistics divided by index of tire production (Table I).

[b] Tire industry payroll index divided by tire-miles produced (Table III).

6 Census of Manufactures. Wages were 13 percent of the value of products in 1925 and 17.5 percent of the value of products in 1935. For the other census years the proportion of labor costs has ranged between these two values.

THE TIRE MANUFACTURING INDUSTRY

B. GAINS TO CONSUMERS

Consumers have undoubtedly received the lion's share of the gains resulting from increased productivity. The change in wholesale prices of tires and in prices per tire-mile are shown in Table XVI.

As may be seen by a comparison of Tables XV and XVI the gains to consumers have exceeded the reduction in labor cost. This is because other elements of cost have been reduced over the period and a large share of these other gains have also accrued to consumers. The most notable illustration of the reduction in other costs has been in the price of crude rubber which has fallen from 65 cents per pound in 1914 to 18 cents per pound in 1940 (April 1st). Tire fabric, the second most important raw material, has also declined in price. Complete data are not available for the period, but the Bureau of Labor Statistics wholesale price index indicates that the price of tire fabric has declined more than 30 percent since 1926.

Thus by the data shown in Table XVI it may be estimated that the consumer got approximately 18 times as many tire-miles for his dollar in 1937 as he did in 1914. A recent survey by the B. F. Goodrich Company revealed that the average motorist's expenditure for tire replacements in 1937 was about $15, whereas in 1917 it was approximately $200. The average annual truck and bus tire bill has been reduced from $600 to $65 within the same period. It is estimated also that cars and trucks were driven almost twice as many miles in 1937 as in 1917.[7] In 1926 the most common size of tire for light cars was the 4.40 x 21. It listed at $23.95 (retail for tire and tube), weighed 15.5 pounds, had a rubber content of 6.1 pounds and gave an average mileage of 14,200. It cost 1.69 mills per mile or $1.55 per pound. The 1938 tire for comparable service is the 6.00 x 16. Its list price (retail for tire and tube) is $19.35 and it weighs 21.9 pounds of which 10.8

---

[7] *The Akron Times-Press*, Akron, Ohio, July 8, 1937. These figures are for annual average tire replacement costs per car. They do not include the cost of tires on new vehicles.

EFFECTS OF INCREASING PRODUCTIVITY

pounds consist of rubber. It gives an average mileage of 26,500 miles and costs only .73 mills per mile and 88.4 cents per pound.[8]

C. GAINS TO THE OWNERS AND MANAGERS

1. The Owners

The gains to the owners for their functions in providing the capital and taking the risks come from the profits of the business. The term profits is used in the customary business sense of the annual excess of income over expenses. The quest for greater profits has been the primary drive responsible for the increase in productivity. Profits are, however, determined by a great many factors other than productivity and labor costs. The elements involved are so complex that it is practically im-

TABLE XVI

WHOLESALE PRICES OF TIRES FOR SELECTED YEARS, 1914–1937

Indexes: 1914 = 100

Year	Wholesale Price Index	
	Per Tire [a]	Per Tire-Mile [b]
1914	100.0	100.0
1919	120.9	90.0
1921	103.5	65.9
1925	56.9	19.9
1929	31.5	7.4
1932	23.7	4.7
1935	26.4	4.9
1937	32.2	5.6

[a] Bureau of Labor Statistics.

[b] Wholesale price per tire index divided by ratio of average mileage in given year to average mileage in 1914. Wholesale prices were used in the absence of an inclusive measure of retail tire prices. Although statistical data are not available it appears likely that this measure somewhat understates the gains to the consumer. Competition among tire dealers has grown more intensive since about 1925 and this development has probably reduced the distributor's margins. See pp. 55–57.

8 Jerome T. Shaw, " Tire Sales Volume Recovers," *The New York Times*, November 13, 1938, p. 26A. These mileage estimates are for first line tires of the latest design. They are, therefore, higher than the average mileage figures given in Table II.

possible to segregate the profits resulting from increased productivity from the profits resulting from other changes in the conditions of business. All that can be done, therefore, is to summarize the profit history of the industry.

The years from 1900 to 1920 were, almost without exception, highly profitable for the tire industry. Dr. David M. Beights states that the returns on the common stock equity of six major tire companies between 1911 and 1920 averaged 24.5 percent per year.[9] Many persons accumulated fortunes from the earnings of tire securities in this period. The demand for tires was increasing so rapidly that the industry had difficulty in meeting it. The great increase in the capacity of the industry which occurred in this period has been indicated in Chapter III. Although much new capital flowed into the industry, many plant expansions were financed by reinvested earnings. Dr. Beights states that between 1911 and 1920 over 52 percent of the profits earned by the six companies studied by him were reinvested in the companies which earned them.[10] Between 1912 and 1920 these six companies increased their capital investment by 500 percent.[11]

The rapid increase in the productivity of labor in this period, which made it possible to give the consumer better and cheaper tires, was undoubtedly one reason for the rapid increase in the demand for tires. Since the exceptional profits of the industry were due largely to the growth of demand, the owners shared substantially in the benefits of increasing productivity. The generally good business conditions of the pre-war and war years were also in large part responsible for the profits of the expanding new industry. Another reason for the high profits was the favorable financing of the industry. Dr. Beights has

9 *Financing American Rubber Manufacturing Companies*, Abstract of Thesis, University of Illinois, Urbana, Illinois, 1932, p. 7.

10 *Ibid.*, pp. 7 and 8.

11 *Ibid.* These companies were the six largest in the industry at the time and their rate of growth was somewhat greater than that of the industry as a whole. Dr. Beights estimates that the entire industry increased its capital investment by not less than 350 percent in this period.

shown that the practice of trading on equity materially increased the earnings on common stock in this period. Between 1911 and 1920 bonds averaged 22.75 percent of the total capital of the six companies studied by Dr. Beights, yet bond interest payments amounted to only 9.2 percent of earnings. Likewise, preferred stocks, most of which were non-participating beyond a fixed return, averaged 32.75 percent of the total capital yet received a significantly smaller percent of total earnings. Moreover, large amounts of money were raised by the industry on short term loans at rates considerably below the earnings of the industry. Altogether, Dr. Beights estimates that the common stockholders received net annual gains of about 12 percent from trading on equity between 1911 and 1920.[12]

This practice, however, led to some disastrous losses in 1920-1921 and the industry has been much more conservative in its financing since that time. During 1919 and the first few months of 1920 the industry was very prosperous. It was operating at full capacity and huge inventories were built up at high prices because of the fear of an impending shortage of raw materials. Several plant expansions were under way, some of which were financed in part by short term borrowings. The collapse of the boom in 1920 caught the industry unawares. The drop in prices of raw materials and finished products, coupled with the precipitous decline of demand, caused huge losses to many producers. The profit history of the industry is shown in Table XVII. Whereas it had averaged 19 percent profits in 1919, in 1920 the rate was only 2.1 percent, and in 1921 a loss of 12.6 percent was shown. These developments forced some of the large tire producers into bankruptcy and reorganization. Although tire production increased again between 1922 and 1929, the industry has been substantially less profitable than manufacturing industry as a whole since 1922. Thus, the data shown in Table XVII indicate an average annual profit of 4.3 percent of net worth in the tire industry

---

12 *Op. cit.*, pp. 4-8. This indicates that nearly half of the returns on common stock in this period were derived from trading on equity. See p. 132.

between 1922 and 1935, while the comparable figure for all manufacturing industry was 7.6 percent.[13] Although in some years, notably 1925 and to a lesser extent 1924, 1926, and 1927, sizable profits were earned, on the whole the industry has not been very profitable for more than a decade and a half. Dividends were maintained pretty well by drawing on accumulated surplus until 1930 but since the depression they have fallen drastically.

Several factors have intervened to prevent the stockholders from receiving any appreciable share of the gains of increasing productivity in the last fifteen years. The heavy burden of fixed charges resulting from the difficulties of 1921 was of major importance. At this time several of the producers were forced to fund their short term loans with securities paying unusually high interest rates and having very strong protective features. The wide fluctuations in the price of crude rubber have also occasioned large inventory losses to the tire industry. This was an important cause of the 1920-1921 debacle and it was intensified by the operation of the rubber restriction plans and the 1929 depression.[14] The recurrent price wars between tire manufacturers have also seriously curtailed profits. The background for this development was laid when Firestone initiated large price cuts in 1921. The other companies soon met the price cuts and the price situation was not brought back to a semblance of order until 1923. The attempt to reach fuller capacity operation has led to such severely competitive bidding for original equipment contracts that a large part of this business has been done with very little profit to the tire manufacturers. The cause of the most serious of the price wars, however, which kept the industry in an uproar from 1926 to 1936,

---

13 A check on the accuracy of the figures relating to the tire industry is provided by the results of an unpublished study of the thirteen tire companies previously mentioned. See p. 86. The study reveals an average annual profit of 4.1 percent of net worth from 1925 to 1935 inclusive.

14 The seven largest tire manufacturers lost more than $83,000,000 on inventory declines between 1926 and 1931. Cross, Earseman, and Lenaerts, *op. cit.*, p. 93.

was the rivalry between Firestone and Goodyear growing out of Goodyear's policy of private brand contracts with Sears, Roebuck.[15] Price competition in the industry was so vigorous

TABLE XVII [a]

PROFITS IN TIRE MANUFACTURING AND IN ALL MANUFACTURING, 1919–1935

	Percentage Profit in	
Year	Tire Manufacturing	All Manufacturing
1919	19.0	13.1
1920	2.1	9.6
1921	−12.6	1.9
1922	3.6	9.1
1923	6.0	10.0
1924	7.3	8.9
1925	18.9	10.7
1926	8.6	11.0
1927	6.9	8.4
1928	1.3	9.8
1929	5.9	13.4
1930	− 4.8	7.1
1931	− 1.7	3.3
1932	− 2.8	0.3
1933	1.6	3.1
1934	2.4	4.3
1935	4.1	6.7

[a] Lloyd G. Reynolds, "Competition in the Rubber-Tire Industry", *American Economic Review*, Vol. XXVII, No. 3 (September, 1938) p. 464, Table III.

In footnote 13 on that page Reynolds describes the data as follows:

"The measure of profits used is the ratio of net income after fixed charges to the total of preferred stock, common stock, surplus and capital reserves. The series used are unfortunately not homogeneous. The data from 1919-28 are taken from Epstein's study of industrial profits; these figures, as Epstein points out, are biased upward by the dominance of large corporations in his sample and by the omission of failures. From 1929-35, the series for all manufacturing are taken from the reports of the National City Bank, and the series for the tire industry are computed from a consolidated income statement of eleven leading tire companies compiled by Standard Statistics; both of these series are biased upward in much the same way as Epstein's series, though it is not possible to determine in which series the bias is greater."

15 See pp. 55-57.

in this period that financial journals labelled it "chaotic", "murderous" and "insane".[16] Many of these conditions have been eliminated or at least greatly modified within the past three or four years. The most recent source of uncertainty with regard to profits is the rise of the new and militant labor union movement. Progress has been made in labor relations within the last two years, however, so that the present outlook for profits appears to be moderately good.

2. *The Managers*

The corporate form of organization has been practically universal in the tire industry since its origin. As the first producers were all small companies with limited financial resources, the principal investors were in most cases the managers of their plants. This situation still obtains in most of the small and medium-sized concerns in the industry. Also in two of the six largest producers (Firestone and General) the original founders or their heirs have retained the major share of both ownership and management. The other major companies and several other medium-sized companies, however, are managed by executives who are not large stockholders but who receive their returns in the form of salaries and bonuses. Moreover, the incomes of some of the executives who are major stockholders and representatives of the controlling interests of their companies depend more upon their salaries and bonuses than upon dividends. The managers are responsible to the owners of the business, however, and it may be assumed that non-owner managements are motivated by the same forces as owner-managements, that is, the quest for profits.

Adequate data concerning the returns of these managerial officials do not exist. The indications are, however, that their salaries are comparable with those of men in similar positions in other industries. Bonus systems are also rather widely used in the industry. The managerial group has shared in the gains

---

16 See Reynolds' comments on these conditions, *op. cit.,* pp. 459-462.

EFFECTS OF INCREASING PRODUCTIVITY   137

from increasing productivity, at least in so far as advancing productivity has increased profits, and at the same time these profits have been distributed in part as bonuses. From the point of view of production economy, the tire industry has enjoyed exceptionally good management. In spite of the high salaries which some of these managers have received, it is not believed that profits have suffered much from a diversion of the earnings to management. On the contrary, it appears that progressive management by cutting production costs and improving the quality of the product has contributed greatly to increasing productivity and protecting profits.

3. Other Salaried Employees

The growth of the tire industry has provided employment for a large number of technical, supervisory and clerical workers, who are not usually classified as either factory workers or managers of the industry. The expeditures of the industry for the salaries of these persons for the past twenty years have averaged more than 10 percent of the value added by manufacture, and such payments represent amounts equal to from 20 to 25 percent of the wage bill for factory workers. The number of salaried officers and employees in the tire industry has varied as follows:[17]

Year	Number
1914	11,952
1919	40,879
1925	14,766
1929	12,981
1935	8,587

[17] Census of Manufactures. The 1914 figure relates to "Rubber Goods, Not Elsewhere Specified" and the 1919 figure to "Rubber Tires, Tubes and Other Rubber Goods". The figures for the other years relate to "Rubber Tires and Inner Tubes". In order to adjust the 1914 and 1919 figures to make them more strictly comparable to those for the industry as now constituted, the 1914 figure should be reduced by 20 percent and the 1919 figure should be decreased 9 percent. This would adjust the figures to tire production plus about 15 percent other products, which has been the average since 1921. See pp. 29-30 and Table V.

In 1914 salaried officers, superintendents and managers numbered 1,238

In so far as the growth of the tire industry has been a result of increased labor productivity, the employment of so large a number of salaried workers may be said to be an indirect effect of the rise in productivity. No very close relationship exists between the salaries paid the vast majority of these workers and the increase in productivity. The average annual salaries in the tire industry correspond to those paid by manufacturing industries as a whole. Apparently this generalization holds true for technical and research men as well as for the mass of clerical employees.[18]

D. GAINS TO WAGE-EARNERS

1. Wages

One of the more notable characteristics of the tire industry is that since its beginnings it has paid higher wages than most other manufacturing industries. As early as 1909 the full-time annual earnings of the branch of the rubber industry including tires, were nearly 3 percent higher than the average for manufacturing industries. Furthermore, the indications are that the differential was larger in the case of tire manufacturing operations than for the other rubber goods production classified with them. As tire production grew to dominate the group and emerged as a separate industry, the differential of full-time annual earnings in the tire industry over the average for all manufacturing industries grew to 7 percent in 1914, 13 percent in 1919, and 15 percent in 1921. Between 1921 and 1933 this differential in favor of the tire industry averaged more than 16.5 percent, and although more recent data are not available the indications are that the differential has increased somewhat and at the present time is nearly 20 percent.[19]

or 10.3 percent of the total salaried employees. This same group numbered 4,015 or 9.8 percent of the total in 1919. The later Censuses have not separated the two groups.

18 See Drucker, *op. cit.*, pp. 68-70.

19 The latest data available show a 19 percent differential for 1933. These comparisons of average full-time annual earnings between the tire industry

EFFECTS OF INCREASING PRODUCTIVITY 139

As may be seen from Table XVIII, average hourly earning rose from 27 cents in 1914 to 69 cents in the boom year of 1920. In the post-war depression of 1921-1922, they dropped to 58 cents but they rose again to 65 cents by 1923. From that time they rose slowly to 69 cents in 1929 and dropped only to 62 cents in 1932. Since 1932 they have risen continuously, slowly at first but more rapidly since 1933, until in 1938 the average hourly earnings of all employees in the industry were no less than 95 cents. Caution must be exercised in the interpretation of these figures. Reasonably accurate published data have been available only since 1932, and the figures prior to that date are based upon samples.

Wage payment in the industry has long been largely on an incentive basis and many companies use highly complicated wage systems. Some companies also maintain bonus systems based upon production or service records. The piece-rates are usually set after careful time studies in the course of which a base rate is determined for each job. The base rate is one which the average good operator is expected to earn per hour. Employees who fail to average this standard rate after the learning period are discharged or shifted to easier jobs. The workers are encouraged to exceed the standard production, however, and many do so consistently, thereby increasing their earnings. Adjustments in individual rates are continually being made to eliminate inconsistencies in base rates between jobs. Technological changes leading to changes in job requirements are also nearly always accompanied by such adjustments. Considerable periods of time may elapse between general changes in the

and all manufacturing industries were derived from the Census of Manufactures by dividing total annual payrolls by average annual employment for the years given. See Table XVIII, Column 3 and Footnote 4. Also see Cross, Earseman, and Lenaerts, *op. cit.*, p. 72; and E. G. Holt, *Rubber Industry Annual Employment, Wages and Salaries*, and *Annual Employment, Wages and Salaries, and Productivity per Wage-Earner in Different Sections of the Rubber Industry*, United States Bureau of Foreign and Domestic Commerce, Circulars Ru-3479 and Ru-3483, Washington, 1933 (Mimeographed).

TABLE XVIII
WAGES IN THE TIRE INDUSTRY, FOR SELECTED YEARS, 1914-1938 [a]

Year	Actual Earnings				Indexes: 1925 = 100			
	Average Hourly Earnings [b]	Average Weekly Earnings [c]	Average Full-Time Annual Earnings [d]	Average Hourly Earnings	Average Weekly Earnings	Average Full-Time Annual Earnings	Average Real Weekly Earnings [e]
1914	$0.27	$12.90	$ 625.	41	44	42	81
1919	—	26.20	1310.	—	89	88	91
1920	.69	29.20	—	105	99	—	85
1921	.59	26.70	1355.	89	91	92	88
1922	.58	26.50	—	88	90	—	96
1923	.65	29.30	1470.	98	99	99	101
1924	.66	29.60	1500.	100	100	101	103
1925	.66	29.50	1480.	100	100	100	100
1926	.68	30.20	1510.	103	102	102	101
1927	.68	30.50	1535.	103	103	104	105
1928	.68	31.50	1575.	103	107	105	107
1929	.69	30.70	1525.	104	104	103	106
1930	.68	28.90	1460.	103	98	99	103
1931	.65	25.30	1285.	98	86	87	100
1932	.62	20.50	1020.	94	69	69	87
1933	.64	20.55	1035.	97	70	70	93
1934	.78	23.65	1200.	118	80	81	104
1935	.84	27.00	1370.	127	92	93	116
1936	.87	30.70	1545.	132	104	104	127
1937	.95	30.10	1500.	144	102	101	119
1938	.95	28.10	—	—	—	—	—

Footnotes on following page.

EFFECTS OF INCREASING PRODUCTIVITY 141

base rates, however. The adjustments in the rates of individuals or small groups usually exhibit an upward tendency in boom times and a downward tendency in times of depression. The average hourly earnings may change somewhat as a result of such adjustments but the major movements of average hourly earnings are dependent upon general changes in basic rates.

Prior to 1914 the shift from the hourly to the piece-rate method of payment was just beginning, but in the period between 1914 and 1919 most of the plants went over to the piece-

^a Comprehensive wage data relating to the tire industry are available only since 1932 and the data for the earlier years represent estimates derived from a number of sources. The major sources are indicated in the footnotes below but a number of other sample studies have been utilized in adjusting the data to comparable form. The data relate to all wage-earners both male and female in the tire manufacturing industry as defined by the Census of Manufactures since 1921. The data prior to 1921 relate to the Census classification of the industry for the years given and they differ from those after 1921 in that a larger proportion of non-tire products was included prior to 1921. See Chapter II, pp. 29-30 for a further discussion of the differences between the Census classifications of the industry before and after 1921.

^b 1914–1931, M. Ada Beney, *Wages, Hours and Employment in the United States, 1914–1936*, National Industrial Conference Board, Study No. 229, New York, 1936, Table 28, pp. 148-151. The N.I.C.B. data relate to the rubber industry as a whole, although the sample was heavily weighted with tire factories. They have been adjusted to conform to the Bureau of Labor Statistics data on the basis of the percentage differentials between the two series since 1932. Data for the period 1932-1938 were taken from the United States Bureau of Labor Statistics, *Employment and Payrolls*.

^c 1914–1922, Beney, *op. cit.*, with adjustments similar to those indicated in footnote b. 1923-1938 data from Bureau of Labor Statistics, *Employment and Payrolls*.

^d 1914-1921, Census of Manufactures. 1923-1937, Bureau of Labor Statistics, *Employment and Payrolls*. Total annual payrolls were divided by average annual employment. Such data are difficult to interpret, being frequently quoted as measures of average annual earnings, but a more accurate interpretation is that they represent the average annual earnings of workers who received the maximum number of weeks of employment with the average number of hours per week for each year. Such an interpretation requires that they be termed the average annual earnings per man-year or average full-time annual earnings.

^e Index of average weekly earnings divided by the index of the cost of living of industrial wage-earners in Cleveland, Ohio, both indexes being on a 1925 base. The index of the cost of living was secured from the United States Bureau of Labor Statistics, Cost of Living Division. The Cleveland index was deemed to be most representative for the tire industry because sample studies have shown Akron prices to correspond closely with those of Cleveland. The use of the national cost of living figures would have made only relatively slight changes in the resulting figures for real wages, however.

rate basis for production workers and practically all of the plants were on this basis by 1929.[20] The change in method of wage payment was in most cases accompanied by an increase in average hourly earnings. Prior to 1914 the rates were rising slowly but the period from 1914 to 1920 brought a series of general increases as a result of the great war. Between 1920 and 1922 a series of decreases in general wage rates reduced hourly earnings by about 16 percent, but they rebounded in 1923 to an average of 65 cents which was less than 6 percent below the war-time peak. Between 1923 and 1930 there were practically no general changes in basic wage rates in the industry. The slowly rising trend between 1923 and 1929 was probably due to: (1) increased individual effort and efficiency resulting in a large number of workers exceeding the base rates; (2) the increasing concentration of the industry in the high wage plants and areas, especially in Akron; and perhaps also (3) a larger proportion of reports from the tire establishments and departments as distinguished from those producing other rubber products.[21]

Average hourly earnings sagged very slightly in 1930, but in 1931 and 1932 the depressed condition of the industry forced general base rate reductions. In spite of this the average hourly

[20] In most plants about 80 percent of the workers are now on piece rates. None of the managers interviewed by the writer reported any significant change in this proportion in the last ten years.

[21] See *Wages and Hours of Labor in the Automobile Tire Industry, 1923*, Bureau of Labor Statistics, Bulletin 358, pp. 1-10; and A. F. Hinrichs, *Akron Rubber Report*, Memorandum from the Chief Economist of the United States Bureau of Labor Statistics to John R. Steelman, Director of the Conciliation Service, April 27, 1938, Washington, pp. 8 and 28 (Unpublished Manuscript). A summary of the latter report appeared in the *Akron Times-Press* and the *Akron Beacon-Journal*, Akron, Ohio, on May 14, 1938.

The B. L. S. study shows average hourly earnings of 72 cents in tire manufacturing in 1923, whereas the figure given for this year in Table XVIII was 65 cents. The difference is mainly due to the fact that persons who were employed in the tire industry but were working on non-tire products were excluded from the Bureau's sample but included in the samples of the present author. As a consequence, the proportion of females and of certain other groups of lower paid workers was somewhat smaller in the Bureau's study than in the present one.

EFFECTS OF INCREASING PRODUCTIVITY   143

earnings declined only slightly more than 10 percent between 1929 and 1932. It is suggested that one reason for the relatively small decline in average hourly earnings in this period was that many workers earned considerably more than their base rates per hour. In many cases where the ordinary course of operations called for minor readjustments downward these changes were not made because the unusually short hours had greatly reduced weekly earnings. It is also probably true that the short hours caused many workers to speed up to the limit of their endurance in the effort to increase their total earnings.[22] Since 1933 several base rate increases have combined with continued increasing efficiency to bring average hourly earnings to the 1938 peak of 95 cents.

Higher hourly earnings by no means always bring with them the higher weekly or annual earnings which are of much greater importance to the workers. As may be seen by comparing columns 4 and 5 of Table XVIII, the trend of average weekly earnings corresponded fairly closely with that of average hourly earnings between 1914 and 1929. During this time the decline in average hours per week was too gradual to have much effect on weekly earnings. Between 1929 and 1932, however, average weekly earnings dropped 33 percent while average hourly earnings declined barely 10 percent. Although the percent of increase in average weekly earnings between 1932 and 1937 was approximately the same as in average hourly earnings (47 percent in weekly earnings and 53 percent in hourly earnings), in 1937 average weekly earnings were 2 percent below 1929 whereas average hourly earnings were 37 percent above that level. The difference between the two measures was, of course, due to the sharp decline in average hours of work per week since 1929. Average full-time annual earnings have closely paralleled average weekly earnings. They rose from about $625 in 1914 to $1310 in 1919 and continued in a general upward direction to $1575 in 1928.[23] After a drop to $1020

[22] Croxton, Lapp and Hanna, *op. cit.*, pp. 28 and 31.
[23] On the basis of the close correlation between average full-time annual

in 1932 they rose steadily again and stood at about $1500 in 1937.

Since weekly earnings are the most accurate of the three series, real earnings have been computed by adjusting weekly earnings by the index of the cost of living. It is perhaps somewhat surprising that the 138 percent increase in the money rates of weekly earnings in the period between 1914 and 1929 resulted in an increase in real earnings of only 30 percent. The greater part of this gain was reaped in the period between 1914 and 1923 as a result of the war boom, but real earnings rose slowly between 1923 and 1929, keeping pace with the gradual rise of money earnings. Most of this gain was lost in the period 1929 to 1932 but since that date real weekly earnings have climbed to new peaks, considerably above any previous levels. Thus in the peak year of the nineteen-twenties, 1928, real earnings were only 7 percent above the 1925 level, although at that time they were higher than in any previously recorded peak; but by 1936 they were 27 percent above the 1925 level and 19 percent above the 1928 peak. Whether or not such large gains are to be maintained remains to be seen, however, as in 1937 they declined to only 19 percent above 1925, or 11 percent above 1928.

Although other factors may have a considerable influence in the short run, wages normally reflect changes in productivity. It is suggested that the principal reason why the tire industry has been able to pay higher wages than most manufacturing industries is that the productivity of labor has been higher in the tire industry. Likewise, the marked upward trend of wages in the tire industry is largely explainable in terms of increasing labor productivity. Other factors which affect wages do so within the limits marked out by labor productivity. General business conditions may increase or decrease the lag between an increase in productivity and an increase in wages, but wages and productivity usually move in the same direction.

earnings and average weekly earnings, it is estimated that the former figure was approximately $1460 in 1920 and $1325 in 1922.

It should be noted again that the productivity measures used in this study represent changes in physical output per unit of labor *over a period of time*—that is changes in average productivity. As such they have no direct connection with the theory of marginal productivity by which wage rates or the proportions of the factors employed *at any given moment of time* are determined.[24] Thus all that we may conclude is that an increase in the average productivity of labor in the industry over a period of time tends to raise the marginal productivity of workers in the industry and so tends to raise wages. The piece-rate method of wage payment in the tire industry tends to reduce the lag somewhat. Although individual rates are usually readjusted after a significant technological change in an operation, the new rates may give the workers some of the gains resulting from the change. Workers not thus directly affected receive their gains from the general wage increases which come from time to time, largely as a result of increasing productivity.

2. *Wage Differentials*

In connection with the above description of wage trends it should be noted that the industry is marked by large wage differentials. Thus in 1923 the Bureau of Labor Statistics reported the average hourly earnings in tire manufacturing as $0.74 for males and $0.46 for females, a differential of more than 60 percent.[25] These differentials remained between 50 and 60 percent between 1923 and 1929, and in 1938 male average hourly earnings were slightly more than 50 percent higher than those for females.[26] The differences between male and female

---

24 See Chapter I, pp. 22 and 23. The theoretical doctrine of marginal productivity is a static analysis, which assumes that the volume of capital equipment is constant and that there are no changes in technology.

25 *Wages and Hours of Labor in the Automobile Tire Industry, 1923*, *op. cit.*, p. 3.

26 *Rates of Wages, Fluctuation of Employment, Wage and Salary Payments in Ohio, 1923*, Bulletin of the Department of Industrial Relations and the Industrial Commission of Ohio, Division of Labor Statistics, Re-

average hourly earnings are not, however, arbitrary discriminations on the basis of sex. They result mainly from the circumstance that most females in the industry are employed on unskilled or semi-skilled jobs whereas the more highly skilled jobs are filled by men. On the jobs on which both men and women work, the basic piece-rates are usually the same for both but the men frequently are able to achieve higher hourly earnings because of greater output. The wage rate differentials between the highest and lowest paid jobs within the same plants frequently exceed 100 percent.

Large geographical differentials in wages also exist in the industry, the wages being considerably higher in Ohio and Michigan than in other plants scattered throughout the country. Further investigation, however, reveals that wage rates are more closely related to the size of the establishment than to its geographical location. The high wages in Ohio are in large measure due to the location of three large factories in Akron. The wage rates of the other large corporation, the main tire plant of which is located in Detroit, correspond closely with these three. The wages paid in these four plants are considerably above those for the rest of the industry in the country at large, but the other smaller tire factories in the Akron area are forced to pay wages which are very nearly as high as those paid in the large tire factories nearby. The wages paid by the branch factories of the large Akron and Detroit concerns, which are located elsewhere are below those paid in Akron and Detroit. Tire factories are usually among the highest paying manufacturing industries in the communities in which they are located.[27]

port No. 8, Columbus, Ohio, 1924, pp. 22 and 165; same title, Report No. 19, 1928, pp. 15, 96, and 192; and same title, Report No. 26, 1929, pp. 22 and 125; Hinrichs, *op. cit.*, pp. 10, 14, and 15.

[27] In support of the above contention that wage rates vary more closely with the size of the establishments than with the geographical location, Hinrichs' classification of tire establishments on the basis of wages may be cited. Dr. Hinrichs divided the industry into three main groups: (1) Ohio-Michigan; (2) California; and (3) "outside area" (including plants

EFFECTS OF INCREASING PRODUCTIVITY   147

3. Hours of Labor

The wage-earners in the industry receive benefits from productivity increases when they lead to reductions in the hours of work. Data concerning the average weekly hours worked in the industry have been introduced in Table VIII, Chapter IV above. Between 1914 and 1929 the average hours per week were gradually reduced from slightly in excess of 49 to slightly less than 45. The eight-hour day with the five and one-half (and in some few cases the five) day week had become standard by 1929. Within the next two or three years hours declined drastically, principally because of the policy of spreading the work. Since 1932 average hours per week in the industry have fluctuated between 30 and 36. The six-hour day is standard in many plants in the industry (including most of the larger ones) and workers' organizations are seeking to make it the universal standard.

The benefits the workers have derived from shorter hours cannot be realistically appraised except in combination with wages. The trends in these directions may be summarized by saying that in 1937 the average tire worker worked 35 percent fewer hours than in 1914 but earned 140 percent more money or 47 percent more real buying power. This result was made possible by an increase of average hourly earnings from 27 cents to 95 cents (a 252 percent increase) during the same period. Clearly the reduction in hours in combination with an increase in earnings was made possible by increasing pro-

---

located in Alabama, Colorado, Connecticut, Illinois, Indiana, Maryland, Massachusetts, New York, New Jersey, Pennsylvania and Wisconsin). The Ohio-Michigan area, which pays much the highest rates, is clearly dominated by three large Akron producers and one large Detroit plant. The California area which pays somewhat lower wages is dominated by medium-sized branch plants of these four producers. The "outside area", in which wages are much lower, includes most of the rest of the industry. The "Outside Area" is not a very homogeneous group since there are three or four plants in every wage class between 90 to 95 cents per hour and 60 to 65 cents per hour. The significant point is that these groupings are not along geographical lines. Hinrichs, *op. cit.*, pp. 13 and 23-29. See also Chapter VIII in the present volume.

ductivity. Thus the wage-earners who have remained employed in the tire industry have received substantial benefits from the productivity advance.

CHAPTER VIII

GEOGRAPHICAL SHIFTING OF THE INDUSTRY

WITHIN recent years shifts in the geographical distribution of tire manufacturing establishments have attracted widespread attention. Some of these developments are so new that they have not yet been interpreted in terms of the history of the industry. Although they are by no means entirely attributable to productivity trends, their relationship to productivity is undoubtedly close. This chapter is therefore devoted to a historical survey of geographic movements of the industry. It is believed that these changes are so important that this study of labor productivity would be incomplete without an appraisal of them.

THE FIRST LOCATION OF THE INDUSTRY

The first company to enter the business was the Roxbury India Rubber Company which was founded in Roxbury, Massachusetts in 1832. Several other rubber factories were started in this region within the next two or three years but all of them failed because of the unsatisfactory nature of their products. Charles Goodyear, who had been carrying on his experiments with rubber in New Haven, Connecticut, after years of failure made his momentous discovery of the vulcanization process in 1839. Thus the basis was laid for the modern rubber manufacturing industries. Soon a number of companies were formed to manufacture rubber goods under Goodyear's licenses. The pioneer factories were in Connecticut, Massachusetts and Rhode Island. These states continued to hold the lead for some time, although the industry soon spread to New York, New Jersey and Pennsylvania. The eastern states dominated the industry until after the dawn of the twentieth century and they are still the leading producers of certain types of rubber goods, notably rubber footwear, belting and hose.

It was not until the development of the automobile tire got underway that the center of production shifted from the East to the Middle West, and Akron, Ohio became the tire and rubber capital of the nation.

The Beginnings of the Industry in Akron

The first rubber factory to locate in Akron came there largely as a result of the activities of a group of local civic boosters. In the 1860's Dr. Benjamin Franklin Goodrich abandoned his medical practice at Jamestown, New York, and started out in the real estate business with J. P. Morris, an attorney. As a result of one of their real estate deals, they acquired a small rubber manufacturing plant at Hastings-on-the-Hudson, New York. Although this plant soon failed, the partners were so convinced that the industry offered a future that they bought another factory at Melrose, New York. In 1869, just as the eastern competition was proving too great for this small company, the partners happened to see a booster pamphlet from Akron offering substantial advantages to new industries. Goodrich decided to move to Akron and with his brother-in-law, Harvey W. Tew, formed Goodrich, Tew and Company to manufacture rubber goods in Akron. The Melrose plant's machinery was moved to Akron, and in 1869-1870 a small plant was erected there out of $5,000 invested by Morris and $18,000 put up by eighteen civic boosters in Akron. The concern, although it has gone through several reorganizations, has operated continuously since 1870. Since 1879, it has been known as the B. F. Goodrich Company, now one of the nation's leading tire and rubber manufacturers.[1]

Other persons in Akron and vicinity saw its expanding profits and soon several other rubber factories were established there. The Miller Rubber Company had its beginnings in 1892, and The Sherbondy Rubber Company (which became Diamond in 1896) was launched by Ohio Columbus Barber, the founder of the Diamond Match Company, in 1894. Both of these com-

---

1 Wolf and Wolf, *op. cit.*, pp. 401-427.

panies were later merged with Goodrich. The Goodyear Tire and Rubber Company was launched in Akron in 1898 by the Seiberlings, Kelly-Springfield started there in 1900, and the Firestone Tire and Rubber Company entered the field in 1905.[2] A number of other smaller tire and rubber concerns sprang up in the Akron area within the next few years.

In the meantime the rubber industry was also developing elsewhere. In 1892 the United States Rubber Company and the Mechanical Rubber Company were founded in New Jersey. Both companies, but especially the former, grew rapidly both by expansion of their own business and by the absorption of many smaller companies. It is interesting to observe that the latter company absorbed one small Akron Company before the turn of the century. This was the India Rubber Company which was founded by the Seiberlings in Akron in 1896 and sold to the Mechanical Rubber Company in 1898. The company operated for a time in Akron but later the plant burned down and was never rebuilt. Thus one of the companies that later became the " big four " of the industry left Akron before its big boom.[3]

The Mechanical Rubber Company was reorganized to form the Rubber Goods Manufacturing Company in 1899 and the new firm was absorbed by the United States Rubber Company in 1905. At this time the new giant was frequently called " the rubber trust " and was accused of dominating the industry. It had about twenty factories most of which were in the East but there were a few in the rapidly growing Middle West. The company was especially strong in rubber footwear, which was at the time regarded as the industry's most profitable line. It was also strongly entrenched in mechanical rubber goods pro-

[2] The Goodyear Tire and Rubber Company's only connection with the inventor Charles Goodyear, was that its organizers paid the heirs of the great inventor a small sum for the use of his name. It is nevertheless appropriate that the name of the inventor of the process upon which the industry rests should live in the name of the world's largest tire and rubber company.

[3] Wolf and Wolf, op. cit., p. 419. The India Rubber Company mentioned herein is not to be confused with the India Tire and Rubber Company which operated continuously in Akron from about 1900 to 1936.

152   THE TIRE MANUFACTURING INDUSTRY

duction. The Akron companies, therefore, found it more profitable to concentrate on the less competitive products. Goodrich, although making a diverse line of products from the first, began manufacturing solid carriage tires and pneumatic bicycle tires at a relatively early date. Most other Akron companies began making these products immediately after their founding. All of the Akron companies were constantly seeking to develop new products. The fact that Akron in the early 1900's contained a number of small but enterprising rubber manufacturers, already manufacturing considerable quantities of carriage and bicycle tires and actively seeking new products, gave that city a tremendous advantage when the demand for pneumatic tires for automobiles began its spectacular growth.

Centralization in Akron

A. prior to 1914

Only a few pneumatic tires for automobiles were made late in the nineteenth century, but production increased rapidly in the first decade of the twentieth century, especially after the beginnings of mass production of automobiles in 1908. By 1912 the $250,000,000 sales of the rubber industries were divided as follows: [4]

Group	Number of Plants	Capitalization	Sales
Akron Companies [5]	20–25	$113,000,000	$95,000,000
U. S. Rubber Co.	27	120,000,000	91,000,000
Others	250	—	64,000,000

The Akron companies already in 1912 accounted for more than one-third of the sales of the rubber industries in the country. At the time of the foundation of the first rubber factory in Akron in 1870, the city's population was only 10,000, but by 1900 it had risen to 42,000, and in 1910 Akron boasted 69,000 inhabitants.

---

[4] Wolf and Wolf, *op. cit.,* pp. 428-429.

[5] Ten good sized companies and ten or fifteen small ones. Goodrich, Goodyear, and Firestone, ranking in that order, together accounted for 75 percent of the Akron total.

B. 1914-1929

By 1914 the rubber industry was producing over $300,000,000 worth of products per year, and of this total tires represented over $146,000,000 or nearly 49 percent. In that year Ohio included 43 percent of the wage-earners and produced 49 percent of the value of products of the Census division of the rubber industry which included tires. By 1919 Ohio included 53 percent of the wage-earners and 56 percent of the value of products of the corresponding division of the industry. In 1929 Ohio included 66 percent of the wage-earners and produced 65 percent of the value of products of the " Rubber Tires and Inner Tubes " industry.[6] The plants located in Akron and its environs accounted for the greater part of the tire production of Ohio.

The concentration of the automobile industry in Detroit gave the Akron plants a transportation cost advantage over the eastern producers on original equipment orders.[7] Other factors were that Akron was located in the industrially expanding Middle West within an overnight ride of more than 55 percent of the country's population, 69 percent of the industries, 71 percent of the income tax returns, 76 percent of the bank deposits, and 54 percent of the automobile registrations.[8] Moreover, Akron became known as the source of a labor supply exceptionally well skilled in tire manufacturing. Probably the most important factor of all in the concentration of tire manu-

---

[6] Census of Manufactures, 1914, 1919 and 1929. In 1914 tires were included in the " Rubber Goods, Not Elsewhere Specified " division of the rubber industry. In 1919 tires were included in the " Rubber Tires, Tubes and Other Rubber Goods " division of the rubber industry. Since 1921 tire products have been classified as a separate industry. Tire products, however, clearly dominated the groups in which they were included in 1914 and 1919. See Table V above.

[7] The United States Rubber Company soon reversed this advantage, however, by establishing a tire plant in Detroit.

[8] Akron Chamber of Commerce. The figures apply to the 1930 Census. In that year Akron was the 35th city in the United States in population but was 12th in the value of its manufactures.

facturing in Akron, however, was that the main Akron factories led the industry in mass production methods. Although hourly earnings were on the average from 25 to 30 percent higher in Akron than in the industry as a whole, the Akron factories were generally believed to have the lowest costs in the nation. That is to say the productivity differential in favor of Akron more than compensated for the higher wages paid there.[9] In order to analyze these developments more closely this period has been divided into two parts 1914-1920 and 1921-1929. Although the trend of the industry was toward Akron throughout the period there are some significant differences between the two sub-periods.

1. The Expansion of the Akron Companies, 1914-1920

This was a time of bonanza-like prosperity for the tire industry during which the profits were high and the number of factories multiplied. The General Tire and Rubber Company, established in Akron in 1916, was the most important new producer to appear at this time. This company has been so successful in good and bad times alike that it is now a serious contender for fifth place in the industry. It steadily built up its position in the industry by an almost unsurpassed manufacturing efficiency and by its policy of producing super-quality tires which sold at a premium above the first line tires of the larger companies. Most of the new plants were small, however, so that the major part of the growth of the industry came from the expansion of the existing plants, especially those of the leading producers. Thus Dr. Beights shows a three and one-half fold increase in the capitalization of the industry between 1912 and 1920 as compared with a five-fold increase for the six

[9] The wage comparisons were made by using the average hourly earnings data for the industry as given in Table XVIII, Column 1, and the results of a twenty-year study of average hourly earnings in one major Akron company. The latter data were released by James P. Miller, Cleveland Regional Director of The National Labor Relations Board in an article in *The Akron Times-Press*, April 15, 1938. The data for the period 1914-1920 were used for the above comparison.

largest producers. Within this group the two companies outside of Akron, The United States Rubber Company and The Fisk Tire and Rubber Company, increased their respective capitalizations by 100 percent and 700 percent. The increases in the capitalization of the leading Akron concerns for the same period were: Kelly-Springfield 250 percent; Goodrich 250 percent; Firestone 700 percent; and Goodyear 1100 percent.[10] The United States Rubber Company was decentralized with some thirty plants, many of which produced tires only as a sideline. The Akron companies, however, did their expanding adjacent to their original plants and were therefore the first to reap the advantages of large-scale production. They employed the latest machinery and equipment and they were the pioneers in the mechanization and scientific management which so greatly increased the productivity of labor. For these reasons the large Akron companies had the highest labor productivity, the most skilled laborers, and probably also the lowest labor costs in the industry. The Akron companies as a group also had the lowest total costs, paid the highest wages, and earned the greatest profits.

The rapid growth of the tire industry and its concentration in Akron led to a tremendous boom in that city. The population increased from 69,000 in 1910 to 208,000 in 1920, thereby making Akron one of the fastest-growing cities in the country. The number of wage-earners employed in the factories of the city increased from 15,800 in 1910 to 65,000 in 1920.

2. *The 1921-1929 Expansion*

The tire industry was hard hit by the depression of 1920-1921 and production dropped from nearly 33,000,000 tires in 1919 to 27,300,000 in 1921. The revival in production was rapid, however, as in 1922 a new high of nearly 41,000,000 tires was reached. Production continued to expand until in 1928 nearly 76,000,000 tires were produced. Nevertheless,

---

[10] Beights, *op. cit.*, p. 4. The Kelly-Springfield Tire Company was located in Akron until 1920. See p. 160.

for reasons which were outlined in the preceding chapter, the decade was one of relatively low profits for the industry as a whole. The Akron companies, however, took the lead in plant modernization and expansion in this period also, and the trend toward larger and more efficient units of equipment continued without abatement. Indeed some of the biggest gains in mechanization, scientific management, and time and motion analysis came in this decade. Moreover, this was the period of the greatest improvement in the quality of tires and of the increase in their average size. Thus the average mileage per tire increased from approximately 3,000 miles in 1910 to 5,000 in 1920, but by 1929 the figure had been increased to 15,000 miles.

The average hourly earnings of wage-earners in the Akron tire factories continued to range from 25 to 30 percent above those of the industry as a whole during this period also. The Akron tire factories remained the most modern and efficient in the world. Their labor productivity was estimated to have been about 30 percent above the national average, thus enabling them to maintain a favorable competitive position.[11]

The geographical shifts in the tire industry between 1921 and 1929 were as follows:[12]

Year	Number of Establishments			Number of Wage-Earners			
	U.S.	Ohio	Other States	U.S.	Ohio	Other States	Percent in Ohio
1921	178	58	120	55,496	28,882	26,614	51.9
1925	126	44	82	81,640	50,350	31,290	61.7
1929	91	32	59	83,263	55,307	27,956	66.4

There was a large decline in the number of wage-earners employed in the industry in 1921 as compared with the peak of 1919, but a considerable part of it was attributable to the

11 See footnote 9, p. 154. These data relate to the period 1921 to 1929 inclusive.

12 Data from Census of Manufactures. Quoted from A. F. Hinrichs, *op. cit.*, p. 3.

changed Census classification of the industry after 1921. The data indicate, however, that in the recovery to 1925 employment in Ohio increased more than 74 percent while employment in other states increased less than 18 percent. Moreover, from 1925 to 1929 employment in Ohio further increased 10 percent whereas outside of Ohio the number of wage-earners employed in the tire industry declined nearly 11 percent. The net result of the changes between 1921 and 1929 was that employment in Ohio nearly doubled whereas employment outside of Ohio remained nearly stationary. That is to say, practically the entire net increase in employment in the industry between 1921 and 1929 occurred in Ohio. The number of plants manufacturing tires was nearly cut in half in this period. It is notable that Ohio had slightly less than one-third of the total plants in the industry in 1921, although they included more than half of the wage-earners and a somewhat larger proportion of the value of products. In 1929, Ohio had only slightly more than one-third of the plants, but they included two-thirds of the wage-earners and a similar proportion of the value of products. The mortality of the small firms in Ohio was nearly as high as in the country as a whole. The larger firms in Ohio, however, expanded at a rate which more than compensated for the extinction of the smaller concerns in this region, whereas such was not the case in the other tire-producing areas of the country.

The Beginnings of Decentralization

A. THE EXPERIENCE OF THE UNITED STATES RUBBER COMPANY

The first experiment in the production of tires in several plants throughout the nation was that of the United States Rubber Company. The company, originally formed by an amalgamation of smaller companies, absorbed several others in the process of its growth so that at one time it had more than forty separate factories. Although it was one of the earliest producers of tires, this company did not become an important tire producer until relatively late, chiefly because it was already doing a large and profitable business in other lines. It has, however, been one of the "big four" tire producers for more than two decades. In recent years approximately 40 percent of its total output has been in the tire line.

Although it did not expand as rapidly as its Akron competitors between 1900 and 1919, the company nevertheless did a large business in this period. It survived the 1921 depression without undergoing reorganization but became badly overcapitalized in the process because of the necessity of funding a large volume of short term liabilities. In the first post-war decade its manufacturing operations were conducted either at a loss or at a very small profit. It nevertheless continued to pay dividends out of accumulated surplus and the profits from its extensive rubber plantations.[13]

In 1928 when the DuPont estate purchased the controlling interest in the company, it was not a profitable company in spite of the fact that for several years its sales had averaged more than $200,000,000. The several plants of the company were operating much less efficiently and with substantially higher costs than those of its major Akron competitors. The new management decided that its decentralized production was one of the major difficulties and began centralizing operations by products. The tire-manufacturing operations were, accordingly, concentrated chiefly in Detroit, and this plant was modernized to make it the equal of any plant in the industry in efficiency.[14] The company now also operates three other tire plants in Los Angeles, Indianapolis, and Eau Claire, Wisconsin. All of its original equipment and the major part of its replacement business are supplied from the Detroit plant, however. The position of the United States Rubber Company in the in-

---

13 "The United States Rubber Company," *Fortune*, February, 1934, pp. 52-56 and 125-127. In 1927 the gross profits on sales (before depreciation and fixed charges) of this company were only 6 percent, as compared with 15 percent for two of its large Akron competitors. In the same year The United States Rubber Company with sales of $60,000,000 more than those of the Firestone Tire and Rubber Company, made $14,000,000 less net profit.

14 *Ibid.*

GEOGRAPHICAL SHIFTING OF INDUSTRY   159

dustry has been substantially improved in the last ten years. This development is in considerable part due to its centralization program, although other factors, notably plant modernization and the reduction of fixed charges, have also contributed to the result. Thus the first experience of the industry with decentralization must be accounted unfavorable.

B. FOREIGN BRANCH FACTORIES

Although more than 95 percent of all tires manufactured in the United States have been sold here, exports were important for a time. Tire exports first passed the 1,000,000 mark in 1920 and averaged more than 1,500,000 per year between that year and 1926. Between 1927 and 1930 more than two and three-quarters million tires were exported annually, but they have dropped to about 1,000,000 per year since 1931.[15] The decline in exports has not been halted by the increase in automobile registrations outside of the United States. American tire producers were able to increase their exports in the 1920's despite the rise in nationalistic trade barriers but the intensification of this movement after 1929 proved too much for them.

Some large American tire companies have attempted to circumvent these trade barriers by the establishment of branch factories in other countries. In the early 1920's American branch tire factories followed the automobile industry's assembly plants into Canada. A few such plants were also built in other countries in this decade but the movement was given a big impetus by the rise of trade barriers since 1929. At present American tire companies have more than twenty such foreign branches in thirteen countries.[16] These plants have not

15 Holt, Circular Ru-3500, *op. cit.*, p. 14, Table III; and Circular Ru-3544, op. cit., pp. 1-2.

16 " U. S. Share Drops," *Tire Review*, June, 1938, p. 39; Cross, Earseman and Lenaerts, *op. cit.*, p. 31. These factories are in Argentina, Australia, Brazil, Canada, England, France, Japan, Java, Mexico, Spain, Switzerland, Union of South Africa, and Uruguay. Plants in some of these countries have found part of their markets in neighboring countries as well.

only reduced American exports but they have pretty well cut off the prospects for any considerable future market for the American factories in these regions.[17] Most American foreign branch factories have done well, however, and this is the first favorable experience of the industry with decentralization.[18]

C. OTHER DECENTRALIZING MOVEMENTS, 1920-1929

1. Kelly-Springfield

In 1920 the Kelly-Springfield Company, one of the oldest Akron companies and in sixth position in the industry, sought to expand its facilities. Its physical plant was so nearly surrounded by Goodyear that there was very little room for an adjacent expansion. At that time the city of Cumberland, Maryland, faced with the depletion of its nearby coal mines, was offering attractive inducements to new industries. When that city offered Kelly-Springfield a free tract of land, tax exemption for ten years, and $75,000 in cash, the company relocated its entire plant in Cumberland. A total of nearly $40,000,000 was invested in the new plant which for a time employed from 3,500 to 4,000 workmen. In spite of the somewhat unfortunate timing of its expansion, the company weathered the 1921 depression and remained fairly prosperous until the middle twenties. But during the price wars, which began in 1926 and were accentuated after 1929, this company lost in

---

[17] Two foreign firms, Michelin et Cie of France and The Dunlop Tyre and Rubber Company, Ltd. of Great Britain, have established branch plants in the United States. The former company operated a small plant here from 1907 to 1930 but has since abandoned it. The latter company has operated a small factory in Buffalo, New York since 1919. Neither of these plants has ever offered any serious competition to American producers although their parent concerns are the largest tire companies in the world outside of this country.

American imports of tires have always been utterly insignificant. Between 1926 and 1936 they averaged about 12,000 per year (Holt, Circular Ru-3500, op. cit., p. 14).

[18] The value of several foreign branch factories to their parent American concerns has been limited by the restrictions which some governments impose on the transfer of profits out of the country.

competitive position and for several years conducted its operations at a loss. It failed to prosper in Cumberland despite the liberal subsidy and tax exemption it had received. Even a wage scale which was approximately 20 percent below Akron's did not put the company in a favorable position. By 1935 it had dropped to little more than one-third its former size. In that year the Goodyear Tire and Rubber Company bought control of The Kelly-Springfield Company for $5,000,000 plus a small amount of Goodyear stock.

Many reasons may be given for the failure of Kelly-Springfield. In retrospect it is certain that its expansion was ill-timed. Perhaps absentee ownership and unprogressive management were other contributing factors. Nevertheless it is possible that had the company remained in Akron it would have kept abreast of the latest developments in the industry and might not have failed. Several small producers failed in Akron during this time also but none were as large as Kelly-Springfield. At any rate the experience of this company is regarded by many observers as another unfavorable experience with decentralization.[19]

2. *The West Coast Branches*

It was also in the decade of the twenties that the first distinctly successful domestic branch factories of the major tire producers were established. In the period of the industry's rapid expansion a number of small tire manufacturing concerns sprang up on the west coast, particularly in California. These concerns, which numbered more than a dozen by 1919, enjoyed considerably lower transportation costs in the west coast market. They were also able to get their rubber more cheaply because of comparative nearness to the major sources of supply in the Orient. The scale of operations of these con-

---

19 See James S. Jackson, "Decentralization of the Tire Industry," *Akron Beacon-Journal*, Akron, Ohio, November 16 to December 6, 1937. This is a series of sixteen articles comparing the competitive situation of Akron and other tire-producing centers. Article No. 7, November 23, 1937, deals especially with Kelly-Springfield.

cerns was too small to attract the serious attention of the large producers for some time. During the same period, the large tire manufacturers were establishing their own textile mills for the production of tire cord in the effort to get a cheaper and better product. The Goodyear Tire and Rubber Company established such a textile mill in Los Angeles in 1920. After noting the advantages of the small producers for the west coast market it established a branch tire factory there also in 1920. The experiment was extremely successful and soon thereafter Firestone and Goodrich also established branch factories in Los Angeles. The United States Rubber Company acquired as a subsidiary the Sampson Tire and Rubber Company of that city also. Most of the small independent producers of this area were driven out of business by this competition and the depression following 1929. But whereas in 1919 less than one percent of the nation's tires were built in California, by 1929 this state accounted for six percent of the total. Its proportion of the 1938 total is estimated at more than ten percent. Los Angeles now rivals Detroit as a secondary center of tire production. The California branches of the large producers are smaller than the main plants of these companies in Akron and Detroit. But the California plants are so new and modern that they now have a somewhat higher value productivity per manhour than the Ohio-Michigan area.[20]

3. *The Goodyear Branch in Gadsden, Alabama*

In 1926 Goodyear began selling special brand tires to Sears, Roebuck and within the next two years Sears was taking a sizable proportion of Goodyear's total production. When the contract came up for renewal in 1928 Sears insisted that Goodyear build a branch factory in the South. Sears, which bought the tires on a cost-plus basis at the factory door, wished thus to effect savings in freight on its southern business. Accordingly Goodyear built a plant at Gadsden, Alabama, with a capacity of about 5,000 tires per day. When it began operating in 1929,

20 Hinrichs, *op. cit.*, p. 22.

it was acclaimed by the Goodyear company as the most efficient tire factory in existence, in spite of its relatively small size. According to testimony introduced at the Federal Trade Commission hearings in regard to the Goodyear-Sears contract, the freight savings on tires were largely offset by the increased freight on materials. Power costs were as low, if not lower, in Akron. Nevertheless the new plant does have some advantages resulting from its location. It is relatively near Goodyear's southern textile mills and is a satisfactory distributing center for the southeastern market. It has been able to secure somewhat lower property tax rates than in Akron. The major element of saving, however, is in labor costs, inasmuch as wage rates are only about two-thirds as high as in Akron. Between 1929 and 1935 this plant was unprofitable, chiefly because it was built during the period of high building prices and operated at a rate far below its capacity.[21]

Trends Since 1929

A. 1929 to 1935

The tire industry experienced a considerable shrinkage in both production and capacity during these years of depression. The extent of the reduction is indicated in the following summary.[22]

Indexes: 1929 = 100

Year	Production			Capacity
	Tires	Rubber Consumed	Tire Miles	
1929	100	100	100	100
1932	58	64	67	90
1935	70	92	88	74

21 Federal Trade Commission, Docket 2116, *op. cit.*; Letter from P. W. Litchfield, President of Goodyear, to R. E. Wood, President of Sears, Roebuck, December 9, 1930; Jackson, *op. cit.*, Article No. 6; Hamilton *et al.*, *op. cit.*, pp. 108-112.

22 Data are from Tables I, III and VII above.

164   THE TIRE MANUFACTURING INDUSTRY

These developments were accompanied by a decline in the number of plants in the industry from 91 in 1929 to 42 in 1935. No new plants were built and very little plant modernization was undertaken. But productivity continued to increase largely because the least efficient plants, equipment, and workmen were eliminated.

In the spring of 1935 optimism began to return to the industry. The N. R. A., which had at first been welcomed enthusiastically by the industry but which soon fell into disfavor, had just been declared unconstitutional. Business in most industries was experiencing a slow improvement in prices, sales and profits. The prices of both crude rubber and tires were sharing in this general advance and the price warfare in tires was subsiding. Foreign branch plants and other factors had reduced annual exports from about 3,000,000 to about 1,000,000 tires but the loss amounted to only about 3 percent of total annual sales. The new market developing from the equipment of tractors, farm implements, and road-making machinery with pneumatic tires had already offset this loss and gave promise of considerable further expansion.

The competitive situation of Akron also looked good in 1935. Ohio's tire factories weathered the depression somewhat better than those in the rest of the country so that whereas in 1929, 66.4 percent of the total tire wage-earners were in Ohio, the ratio had risen to 68.4 percent by 1935.[23] As of the latter date Akron contained the main plants of the three largest tire producers (Goodyear, Firestone, and Goodrich) as well as several smaller producers.[24] In 1935, 52.9 percent of all pneumatic tire casings in the United States were made by the five producers within the corporate limits of Akron.[25] The United

[23] Hinrichs, *op. cit.*, p. 3.

[24] In total tire sales the four leading producers in the country are Goodyear, Firestone, Goodrich, and U. S. Rubber, in that order. In total sales of all rubber products they are Goodyear, U. S. Rubber, Goodrich, and Firestone (1939).

[25] Hinrichs, *op. cit.*, p. 5. Two or three percent should be added to give the total for the Akron area including plants outside of the city limits. These

States Rubber Company with its main tire plant in Detroit was the fourth member of the " big four ". The latter company after slipping in relative position for years had but recently embarked upon its comeback. Clinging somewhat precariously to fifth position was The Fisk Tire and Rubber Company of Chicopee Falls, Massachusetts. It was the only remaining tire manufacturer of any significance in New England, the former center of the whole rubber industry. This company although once a serious rival of the " big four " had fallen upon evil days. It had shrunk to only moderate size and had lost its original equipment contracts as well as many of its dealers. The General Tire and Rubber Company of Akron was seriously challenging Fisk for fifth place in the industry by 1935. Kelly-Springfield, the traditional holder of sixth place, was absorbed by Goodyear in that year. By 1935 each of the four major producers had established a branch factory in Los Angeles but their combined output was only about 7 percent of the nation's total. A considerable portion of it had been diverted from independent west coast producers rather than from the main plants of these companies. The only other branch tire plants of the Akron companies were the Gadsden plant and the newly acquired Kelly-Springfield plant at Cumberland, both owned by Goodyear. Decentralization did not at this time appear to menace Akron's continued dominance in the industry.

Average hourly wages had been rising in the tire industry since 1932, and in 1935 they surpassed all previous levels. The industry had long been characterized by considerable wage differentials between the different producing areas. Akron was the established leader in this respect also, although the hourly wage rates paid in the Detroit plant of the United States Rubber Company were on a par with those in Akron in 1935. For a number of years the Ohio-Michigan average had been

figures somewhat understate the importance of Akron in the industry, however, since the Akron plants make a larger proportion of the bigger and more expensive types of tires.

about one-tenth higher than in California and about one-third higher than the rest of the country excluding California.[26] The wage differential between the Ohio-Michigan and California areas has never been large enough to affect seriously the location of plants in the industry. The principal advantage of the California plants is their lower transportation costs to the west coast market, which is restricted within rather definite limits. The problem of competition on the basis of labor costs, therefore, involves comparisons between the Ohio-Michigan region and the outside area.

Prior to 1935 the productivity differential in favor of the Ohio-Michigan area apparently more than offset the higher wages paid in the latter area. It is indeed reported that the Akron Companies had the lowest labor costs in the industry at the same time that they were paying the highest wages. Consequently it was not thought that the competitive situation of Akron would be seriously impaired when another wage increase was made there early in 1935. The wage differential of the Ohio-Michigan area over the outside area was thereby increased to 27 cents per hour, or more than 43 percent.[27] The Akron plants were accustomed to taking the lead in wage increases in the industry and it was expected that the plants in the outside area would follow and soon reduce the differential to the customary level. As will be seen in the next section, the fact that this did not happen was one reason for a considerable decline in Akron's proportion of the industry.

B. THE NEW MOVEMENT TOWARD DECENTRALIZATION SINCE 1935

Few observers of the industry in January, 1935, would have been able to predict that it was on the verge of the greatest

---

26 Hinrichs, *op. cit.*, p. 23. Following Hinrichs' classification of wage areas they will be discussed as Ohio-Michigan, California, and "outside area." The latter includes all tire plants in the United States except those in Ohio, Michigan and California. See footnote 27, pp. 146-147. The Akron rates were slightly higher than the Ohio-Michigan average.

27 *Ibid.*, p. 24. The Ohio-Michigan average was 89 cents per hour and that of the outside area was 62 cents.

decentralizing movement in its history, or that by 1938 Akron would have lost nearly as large a proportion of the industry as it gained in the previous decade and a half of concentration. The following table illustrates the trend of the proportion of wage-earners in Ohio.[28]

Month or Year	Number of Wage-Earners			Percent in Ohio
	United States	Ohio	Other States	
Average 1935	57,128	39,063	18,065	68.4
February 1937	68,550	43,550	25,000	63.5
February 1938	50,650	29,750	20,900	58.7

From 1935 to February 1937 employment in the industry increased by 11,400 or 20 percent, but the increase amounted to only 4,500 or 11.5 percent in Ohio whereas the increase in other states amounted to nearly 7,000 or 38.4 percent. From February, 1937 to February, 1938, there was a decline of 17,900 or 26.1 percent in employment in the industry. Ohio lost 13,800 or 31.7 percent while other states lost only 4,100 employees or 16.4 percent. During 1937 man-hours declined by 57.5 percent in Ohio and only 35.1 percent in other states. Dr. Hinrichs estimates that in February, 1938, only 55 percent of the total man-hours of the tire industry were in Ohio.[29]

A part of the loss of tire industry jobs in Ohio has been due to the improved position of major competitors elsewhere. The United States Rubber Company in particular has made rapid strides in the last few years. It has modernized its Detroit plant and increased the productivity of labor until it now probably surpasses most Akron producers in efficiency. This company uses the new merry-go-round process for building tires by assembly line techniques, which is not used in any Akron factory. The plant is almost completely unionized and pays wages which are about the same as those paid in Akron. It enjoys unusually good relations with the union and has had

28 Hinrichs, *op. cit.*, p. 4.
29 *Ibid.*, pp. 4-5.

very little trouble with strikes, resistance to increasing productivity, or restriction of production by the workers. In part the excellent labor relations of this company are due to the recognition of the union at an early date without resort to a National Labor Relations Board election. The company was also the first major tire producer to sign an agreement with the United Rubber Workers union, and by giving the union an exclusive bargaining contract it has succeeded in winning the confidence of both the union leadership and the rank and file in a program of cooperation. Today this company is the largest producer of original equipment tires and is in addition the principal supplier of three large mass distributors of replacement tires.[30]

In December, 1939 the United States Rubber Company purchased the business and assets of the Fisk Rubber Corporation. On January 22, 1940 F. B. Davis, Jr., President of the United States Rubber Company, reaffirmed his company's intention to continue the operation of the Fisk plants, " probably at even higher capacity." [31]

Sears, Roebuck, one of the largest mass distributors of replacement tires, since the abrogation of its contract with Goodyear in 1936, has again been buying its tires from several small and moderate-sized companies. One of these suppliers, The Armstrong Rubber Company of New Haven, Conn., which heretofore has not made many tires, is now building a new branch plant in Natchez, Miss., in order to make tires for Sears, Roebuck. The capacity of this plant will be about 2,000 tires and tubes per day. Contracts with Sears have aided several other small companies outside of Akron.

---

30 Montgomery Ward and Company, the Atlas Supply Company, and the Western Auto Supply Company. Statement by Mr. T. G. Graham, Vice-President of the B. F. Goodrich Company in the *Akron Times-Press*, March 5, 1938; see also *Fortune*, February, 1934, p. 125.

31 *Standard Corporation Records*, Standard Statistics Company, Volume XVIII, No. 56, Sec. 6 (March, 1940), p. 18; and Volume XVIII, No. 60, Sec. 6 (March, 1940), p. 61.

GEOGRAPHICAL SHIFTING OF INDUSTRY   169

A more serious recent loss to the Akron companies, however, results from the new ultra-modern tire factory which Henry Ford has built near his River Rouge plant in Dearborn, Michigan. This new plant, constructed at a cost of $5,600,000, is now producing 4,000 tires per day. Ford's plans are said to call for a production of 6,000 tires and tubes per day in the near future and eventual production as high as three times this number.[32] This development is a blow to Akron which promises to become increasingly serious as Ford tire production is increased. Firestone in Akron has long been a large supplier of Ford tires. One whole plant of this company in Akron has in the past been devoted mainly to this purpose.[33]

The above developments have, however, been less important in the loss of tire production to Akron than the establishment of branch factories of the Akron companies elsewhere and the increase in the output of the existing branch factories. The list of such new branch tire factories and expansions of older ones is as follows:[34]

Company	Plant Location	Date	Capacity
	Plant Expansions		
Goodyear	Gadsden, Ala.	1936	5,000
Goodyear	Cumberland, Md.	1936	4,000
	New Plants		
Goodyear	Jackson, Mich.	1937	6,500
Goodrich	Oaks, Pa.	1937	5,000
Firestone	Memphis, Tenn.	1937	10,000

32 E. V. Osberg, "The New Ford Tire Plant," *India Rubber World*, June 1, 1938, pp. 53-64; "Ford Tire Plant," *The Rubber Age*, June, 1938, p. 168.

33 Wolf and Wolf, *op. cit.*, p. 476. Other Akron plants have had a considerable share in the Ford business also.

34 Much of the material contained herein was obtained from interviews by the writer. The capacity figures are minimum estimates. The capacity data for the two expansions relate to the recent increases in capacity not to the total capacities of these plants. Significant employment data are not available as some of the plants are just beginning to operate. It is estimated, however, that from 3,000 to 4,000 are now employed as a result of these expansions and new plants, and that the number may be expected at least

In addition to the above new plants Akron has lost heavily in the shifting of mechanical and miscellaneous rubber goods divisions of the Akron companies elsewhere. Labor costs represent a larger proportion of the total costs of these products than of tires and the average wages paid in factories producing these goods elsewhere are only about 60 percent of the Akron levels. The production of these articles requires less skilled labor than tire production, value productivity is lower, and a larger proportion of women are employed.[35] Among the recent shifts of mechanical and other rubber goods production from Akron are the following:

New Non-Tire Branch Plants of Akron Companies [36]

Company	Location	Year	Products
Goodrich [37]	Watertown, Conn.	1930	Boots and shoes
Goodyear	Windsor, Vt.	1936	Heels and soles
Goodrich	Cadillac, Mich.	1937	Automobile parts (rubber)
General [38]	Wabash, Ind.	1937	Mechanical goods
Firestone [38]	Fall River, Mass.	1937	Latex products
Firestone	Noblesville, Ind.	1937	Mechanical goods
Goodrich	Clarksville, Ky.	1939	Mechanical goods

to double when these plants swing into full operation. See also *The Akron Beacon-Journal*, Articles on "Decentralization of the Tire Industry" by Special Correspondent James S. Jackson, November 16 to December 6, 1937.

35 Hinrichs states that in March, 1938, the B. F. Goodrich Company in Akron paid average hourly wages of 98.6 cents in mechanical rubber goods production. The national average of 150 concerns primarily engaged in the production of these products was 59.7 cents in February, 1938. In the latter month the average hourly earnings in tire production were:

    Akron $1.10
    Outside of Akron .... .84

The differential between Akron and other areas was thus larger in mechanical rubber goods than in tires. Hinrichs, *op. cit.*, pp. 11-12.

36 Sources of information same as given for tire branch factories above.

37 This is the Hood Rubber Co. plant which was taken over by Goodrich. Goodrich's footwear production was shifted to Watertown and Hood's tire production was taken to Akron. Akron lost at least twice as many jobs as it gained as a result of the transfer.

38 This production was not shifted from Akron since these plants enable the companies concerned to produce a new line of products. The construction

The exodus of mechanical and other rubber goods production was so substantial that by 1938 Goodrich was the only company producing any considerable proportion of its non-tire products in Akron.[39]

C. REASONS FOR DECENTRALIZATION

Although the experience of the foreign branches and those in California indicates that a certain amount of decentralization is inevitable, until very recently the advantages of centralized mass production have been sufficient to restrict effectively decentralizing tendencies in the tire industry. Since about 1929, however, the technological developments in the industry have been to a large extent in the direction of more individualized units of equipment. These developments have proceeded at such a pace that a modern plant with a capacity of from 5,000 to 10,000 tires per day can have advantages in process alignment, continuous flow of materials, and mass production which allow it to compete on nearly equal terms with plants having several times this capacity. Although this is true only if the smaller plant specializes in a few types of tires, the recent progress of size standardization has made such specialization feasible. Technological developments of this nature have thus facilitated the decentralization of the industry by market areas in order to secure transportation and marketing savings. In this respect the tire industry has merely followed the decentralization of the automobile assembly plants.

The shifting of plant locations because of the above factors is somewhat hazardous, however, because of the rather narrow margin of advantage to be gained in most cases. Thus it is exceedingly unlikely that anything like as large or as rapid a decentralizing movement would have developed were it not for

---

of these plants outside of Akron represents losses to that city, however, in the sense that both companies in the past have done most of their expanding in Akron and would probably have built their new plants there had conditions been as favorable as they have been until recently.

[39] Statement of T. G. Graham, Vice-President of The B. F. Goodrich Company, in the *Akron Times-Press*, March 5, 1938.

the unfavorable position of Akron with regard to labor costs and labor relations. In the following tabulation the relative wage rates of the Ohio-Michigan area and the outside area are compared.[40]

Selected Month	Average Hourly Earnings (Cents per Hour)		Differential in Favor of Ohio-Michigan	
	Ohio-Michigan	Outside Area [41]	Cents per Hour	Percent
Jan. 1935	83.	62.	21	34
Feb. 1935	89.	62.	27	43
Nov. 1935	89.	63.	26	41
May 1936	95.	63.	32	51
Jan. 1937	96.	67.	29	43
Mar. 1937	104.	69.	35	51
Feb. 1938	104.	74.	30	42

Increases in wage rates were initiated in Akron in February, 1935, May, 1936, and February, 1937. These increases were followed in Detroit and Los Angeles so that Akron and Detroit wages remained comparable and Los Angeles maintained a level about 10 percent below them. The increases in wages in the other areas were much slower and smaller, however, so that the Ohio-Michigan differential increased from about 34 percent to from 40 to 50 percent. This was the period of rapid growth of the United Rubber Workers union, especially in Akron, but the union was not directly responsible for the wage increases. The indications are that the Akron companies put through these wage increases in an effort to stem the tide of the growth of unionism.[42]

The growth of the union has strengthened the tendency for the industry to decentralize. In both the 1933-34 and

---

40 Hinrichs, *op. cit.*, pp. 23-26. In February, 1938, Akron rates were 6 cents above the Ohio-Michigan average. See footnote 35, p. 170.

41 All states having tire plants except Ohio, Michigan, and California.

42 Hinrichs, *op. cit.*, pp. 24-25. In most cases the union did not achieve recognition for purposes of collective bargaining until several months after the major wage increases.

1936-37 organizing campaigns the greatest efforts of the union leaders were concentrated in the Akron area. The very existence of the union seemed to depend upon its acceptance here. Likewise the Akron companies were the most vigorous opponents of unionism. As a consequence Akron became the center of the greatest labor struggles in the history of the industry. It was here that the first mass sit-down strikes were staged, and that coercion and violence on the part of the union were most severe. It was here also that the new federal labor laws were most disregarded by the companies, and that anti-union propaganda and intimidation were most rampant. The union emerged victorious in the majority of these contests, and by the end of 1937 most tire manufacturers in the Akron area were dealing with it either formally or informally. By the end of 1938 the union had concluded written agreements with all but one of the Akron tire producers. There have been few strikes or disturbances in Akron since that time, but the concentration of union strength in Akron tends to prevent a readjustment which would give Akron a more favorable competitive situation and thus remains a factor favoring further decentralization. The union seeks to meet this threat by organizing the outside area but its progress in this direction has been slow. Although there are several locals in the plants of the outside competitors of the Akron companies, very little headway had been made in the outside branches of the Akron companies.

Many of the issues involved in the problem of decentralization were brought to the fore by the B. F. Goodrich Company in February, 1938. Mr. T. G. Graham, Vice-President of the company, in a statement to the employees and the public, pointed out that Goodrich was the least decentralized of the large Akron producers and that the company had undertaken very little plant modernization since 1929. He reviewed the recent movement of the industry from Akron, attributing it to the technological developments which have increased the efficiency of small plants and to the increase in Akron's wage differential. He stated that since January, 1936, at Goodrich base rates had increased 14 percent, efficiency had decreased 6 percent, and labor costs had increased 20 percent. He pointed out that in spite of the increase in base rates average annual earnings had decreased 6 percent. These developments, Mr. Graham contended, made it impossible for Goodrich to continue to compete in the industry under existing conditions. Consequently he proposed the following program to the Goodrich employees:

(1) Reduction of wages for men from $1.125 to $0.925 per hour and for women from $0.725 to $0.625 per hour.

(2) The union to give the company assurances and cooperation in using the equipment and facilities of the company as efficiently as possible.

(3) A work week varying from 30 to 40 hours to be substituted for the present one of 30 to 36 hours.

(4) If the first three proposals were accepted by the union the company would spend not less than $1,500,000 in modernizing its Akron plants and would not decentralize its operations further for six months.

(5) If the union rejected the first three proposals the company would reduce its production in Akron from two-thirds to one-third of its total. This would mean a loss of about 5,000 jobs in Goodrich's Akron factories.[43]

Most of the leaders of the United Rubber Workers union and of the Goodrich local of that organization opposed the wage cut while many civic groups in Akron urged its adoption. The issue was widely discussed and the vote was postponed several times. The union leaders came forward with a counter proposal for a 6 percent wage cut. They also sought to get the company to agree not to cut wages outside of Akron and not

---

[43] *The Akron Times-Press* and *The Akron Beacon-Journal*, Akron, Ohio, March 5, 1938. These papers carried accounts of this proposal and the discussions and controversies it called forth almost daily until June 1, 1938. Other issues of especial interest are those of May 14, and May 20-28, 1938.

The investigation and report made by Dr. A. F. Hinrichs, Chief Economist of the United States Bureau of Labor Statistics (*op. cit.*) was a result of the United Rubber Workers' request for additional information before the vote was taken.

to decentralize further for two years. Mr. James P. Miller, National Labor Relations Board Director of the Cleveland District, proposed that the wage cut be accompanied by a plan for a guaranteed annual wage for permanent workers, and for 20 percent of the net earnings of the company to be distributed to the workers each year as a bonus. The company stated that the original terms were the most liberal that could be offered. They pointed out that, although the proposal represented a cut of about 17 percent for men and 13 percent for women, this left wages at the 1936 level, which was well above any previous peak. In May, 1938, the company's proposal was overwhelmingly rejected by the workers. A subsequent agreement which was signed by the company and the union made no mention of any changes in wage rates or decentralization. Goodrich has not proceeded with its plans for modernization of its plants in Akron. A considerable part of its mechanical goods production has been shifted to its new plant in Clarksville, Kentucky, but it has not further decentralized its tire production up to the present writing.

CHAPTER IX
SUMMARY AND CONCLUSIONS

Productivity

PERHAPS the outstanding characteristic of the tire industry from the point of view of the changing productivity of labor is the rapid rate of growth it has undergone. In Tables I, III, and VIII, data indicating the growth of production and employment from 1914, the approximate date of the emergence of tire manufacturing as a separate industry, through 1937 have been presented. It has been emphasized that the changes in the quality of tires during this period have been as important as changes in the quantity produced. Thus the most significant measure of production, tire-miles, has shown nearly a 40-fold increase in the twenty-three year period.

Even after allowance has been made for the large number of persons employed on non-tire products in the early years of the industry, it is clear that the peak of the industry's employment was reached as early as 1919. Employment has declined since this time in spite of the concurrent decline of average hours per week in the industry. It required about 50 percent more employees to manufacture 154 billion tire-miles in 1919 than it did to manufacture 1,108 billion tire-miles in 1937. With the data available it is impossible to determine the trend of employment in the distribution of tires. Renewal tire sales reached a peak of 52,000,000 in 1928, however, whereas they have averaged only about 30,000,000 per year for the period 1936 to 1938. Moreover, the proportion sold by mass distributors has increased from 2 percent in 1922 to 48 percent in 1937. It seems improbable that the increase in employment among the large scale distributors has been great enough to offset the decline in the number of independent tire dealers and their employees.

The productivity data are summarized in Tables XIX and XX. In the former are presented the various measurements of

TABLE XIX

Labor Productivity in Tire Manufacturing for Selected Years, 1914–1937 [a]

Indexes: 1925 = 100

	1914	1919	1921	1925	1929	1932	1935	1937
Tires per Man-Year								
Actual	340	449	647	1,014	1,108	1,178	1,185	1,165
Index	33	44	64	100	109	116	117	115
Tires per Man-Hour								
Actual	.138	—	.293	.456	.505	.683	.734	.735
Index	30	—	65	100	111	150	161	161
Pounds of Crude Rubber per Man-Year								
Actual	—	—	5,167	9,021	10,147	12,178	13,622	14,572
Index	—	—	57	100	113	132	151	162
Pounds of Crude Rubber per Man-Hour								
Actual	—	—	2.36	4.05	4.53	7.49	8.42	9.19
Index	—	—	58	100	112	185	208	227
Tire-Miles per Man-Year								
Actual (thousands)	1,183	1,285	3,230	10,440	16,875	15,574	16,122	23,240
Index	11	12	31	100	162	149	154	223
Tire-Miles per Man-Hour								
Actual	483	—	1,472	4,690	7,530	9,028	9,982	14,660
Index	10	—	31	100	161	162	216	313

[a] Based on data in Table IX. The tires per man-year and tires per man-hour series are based on equivalent tires.

TABLE XX
Unit Labor Requirements in Tire Manufacturing for Selected Years, 1914–1937 [a]

	1914	1919	1921	1925	1929	1932	1935	1937
Man-Years per thousand tires	2.94	2.23	1.55	.99	.88	.85	.84	.86
Man-Hours per tire .	7.22	—	3.41	2.19	1.98	1.46	1.37	1.36
Man-Years per thousand pounds crude rubber	—	—	.19	.11	.10	.08	.08	.07
Man-Hours per pound crude rubber	—	—	.42	.24	.22	.14	.12	.11
Man-Years per million tire-miles ..	.85	.78	.31	.09	.06	.06	.06	.04
Man-Hours per million tire-miles ..	2,077	—	679	214	132	111	64	68

[a] Data based on Table XIX.

the productivity of labor for selected years. The most comprehensive measures of productivity for comparisons over the entire period are in terms of tire-miles per man-hour. In Table XX the data are presented in a somewhat different form, namely in terms of physical unit labor requirements. Thus instead of stating that in 1914, 340 tires were manufactured per man-year and in 1937, 1,165 tires were produced per man-year, it is shown in Table XX that whereas in 1914 it required 2.94 man-years of labor to manufacture 1,000 tires, in 1937 the same number of tires were manufactured with only 0.86 man-years. Similarly it is shown that the production of 1,000,-000 tire-miles required 2,077 man-hours in 1914 but only 68 man-hours in 1937.

Conditions Attending Increasing Productivity

It has been stressed repeatedly in the derivation and discussion of these statistical measures of productivity that they are simply technical devices relating the quantity of goods pro-

SUMMARY AND CONCLUSIONS 179

duced to the amount of labor-time employed, and that the mere statement of the measures implies nothing as to the causes of these movements. Although the factors responsible for these developments are too complex to be weighted numerically, they have been discussed in considerable detail. First of all it has been noted that the rapid upward trend of productivity has been closely associated with the growth of the industry. The enormous expansion of automobile transportation and therefore of the market for tires has been the most vital factor conditioning the increase in productivity. Closely associated with the growth of the physical volume of production and increase in labor productivity has been the improvement of the quality of tires. Moreover, although the data for recent years indicate a slowing down of the growth of physical production and productivity, the quality of the product is still undergoing steady improvement.

Bearing these general factors in mind the more specific factors responsible for the increase in productivity have been classified into two groups, which for want of better names have been designated " mechanical " and " non-mechanical " factors. The " mechanical " factors include the application of mass production techniques to the industry. The capital investment of the industry has been multiplied several-fold during the period studied and this has made possible the substitution of machinery for manual labor on a vast scale. Likewise the industry has witnessed many technological improvements which have increased productivity without requiring additional capital investments. Many illustrations of these developments have been given in Chapter VI above.

The analysis of the preceding chapters indicates that the " non-mechanical " factors have also been of very considerable importance in the increase in productivity in the tire industry. There has been a decided tendency toward concentration of production in the more efficient firms and plants. Thus these organizations grew and applied the newest techniques while the others declined or failed. The superiority of the growing concerns was apparent not only in technical equipment, methods, and processes but also in management. This survey shows clearly that the growth of the scientific management movement in the tire industry has had a significant part in increasing labor productivity. The time and motion study techniques developed by the new school of management have eliminated a great deal of waste in labor utilization by the accurate determination of the easiest and most efficient method of performing each operation involved in tire production. Scientific management has also led to the more widespread use of mass production techniques involving more and more division of labor, which in turn has facilitated greater mechanization. Plant layouts have been improved and the speed of production of the average worker has been increased. The latter change has been accomplished chiefly by the use of incentive methods of wage payment in as wide a range of occupations as possible. These systems of remuneration are also calculated to stimulate the management to maintain an uninterrupted flow of work. The other methods of scientific management have been developed in combination with greatly improved personnel departments charged with the careful selection and training of new workers.

A policy of attempting to maintain the best possible morale and cooperation of the labor force has also been followed. The high wage policy of the tire industry was undoubtedly a very important factor in securing labor's collaboration in many of these productivity stimuli. Since the rise of militant trade unionism in the industry, however, the high wage policy seems to be less effective than formerly as a method of securing labor's cooperation. Labor relations are more difficult now in spite of the fact that hourly earnings have surpassed all previous peaks. Moreover, labor relations are more troublesome in Akron, where the highest wages in the industry are paid, than they are elsewhere. In recent years the problem of labor relations has been foremost in the industry. Some managements have attempted to meet the difficulty by bending every

SUMMARY AND CONCLUSIONS   181

effort to prevent union organization. Others have cooperated with the union leaders in every respect and are operating under union agreements. Some have vacillated between the two policies. None of these policies have been successful in all cases and the resulting unstable conditions in labor relations have been in large measure responsible for the reduced rate of productivity increase since 1933.

Prior to 1929 the work-week was gradually being shortened and the eight-hour day had become standard in the industry. Since that time the work-week has been considerably reduced and the six-hour day is now in effect in many plants. This development facilitated the reduction of waste in labor-time and has contributed to the increase in productivity especially on a per man-hour basis. Since 1930 the concentration of production in the most efficient plants and among the most efficient workmen and units of equipment have been primary factors in increasing productivity. In spite of the fact that the trend toward greater division of labor has on the whole reduced the skill requirements of tire workers, the increased efficiency of the workers has been a factor of considerable importance. This development has come about chiefly through an increase in the average rate of speed at which the workmen are required to operate. The efforts of management in this direction have been supplemented by the workers' own pressure to hold their jobs during a period of declining employment. The attempts to reduce the amount of seasonal fluctuation in production and employment have also to some extent aided in increasing productivity.

Effects of Increasing Productivity

For the analysis of the effects of increasing productivity on employment, the history of the tire industry has been divided into three stages, namely: (1) Prior to 1919; (2) 1920 to 1929; and (3) 1930 to 1937. During the first period the extraordinary growth of employment in the industry was aided and indeed to a large extent made possible by the rapid increase in labor productivity. During the second period productivity increased more rapidly than production so that, although the industry was still growing, there was a net decline of some 17,000 jobs in tire manufacturing. During the third period both production and employment were drastically curtailed as a result of the great depression. By 1937, however, production in terms of tire-miles and rubber consumed had again surpassed all previous levels, but employment in tire manufacturing remained some 22,000 below 1929. Thus, although the development of the tire as a new product and the subsequent increase in the productivity of labor in tire manufacturing had created nearly 90,000 jobs in a new industry by 1919, a continuation of the increase in productivity has led to a reduction of more than 39,000 jobs in the industry since that time. The increased productivity of labor in the tire industry at the same time that it was making possible the production of tires with less labor, was also creating a problem of technological unemployment.

Of course, all of the increase in productivity does not represent a net gain in pure efficiency, since it was accomplished in part by the greater use of capital and resources. Although the present development of knowledge and techniques does not make it possible to determine the net gain in efficiency of economic production by reason of the productivity increase, it is obvious that the greater part of the increase in productivity does represent a real net gain in the efficiency of production. A rough appraisal of the amount of this net gain may be found in the historical analysis of the gains accruing to the several groups which were in a position to benefit from increased productivity in tire manufacturing.

The wage-earners employed in the tire industry have reaped large gains. In 1937 the typical tire worker, although averaging 35 percent fewer hours than in 1914, earned about 140 percent more money or 47 percent more real buying power. In the same period average hourly earnings rose from 27 cents to 95 cents, a 252 percent increase. Full-time annual earnings in the tire industry in 1914 were approximately 7 percent above those in all manufacturing industry, and this differential has steadily increased so that it is now about 20 percent. Although other factors such as the rise of the standard of living and the emergence of collective bargaining have contributed to these wage increases, it must be emphasized that it is impossible to have any sustained upward trend in wages unless there is a similar trend in productivity.

Wage-earners have not received all or even the major share of the gains from increasing productivity, however. The total of the gains made possible to other groups is revealed in the declines in labor costs. Between 1914 and 1937 the labor cost per tire declined nearly 30 percent and the labor cost per tire-mile declined by more than 87 percent. That the consumers have received the greater part of this gain is indicated by the fact that within the same period the wholesale price per tire declined 68 percent and the price per tire-mile declined by more than 94 percent. That is to say, the 1937 consumer got approximately 18 times as many tire-miles per dollar as the consumer of 1914. These benefits to consumers have exceeded the gains made available by increasing productivity because other elements of cost per tire-mile (notably the cost of materials) have also declined during the interval.

The owners and managers of the industry reaped huge profits and large bonuses in the exceedingly prosperous years of the tire industry's early growth. From the point of view of profits, however, this period of the industry's history ended about 1919. The annual profit between 1922 and 1935 averaged 4.3 percent of net worth in the tire industry as compared with 7.6 percent in all manufacturing industry. Factors entirely apart from productivity changes, such as excessive fixed charges and other unwise financing practices, poor timing of expansion, and ruinous price wars appear to have been chiefly responsible for the failure of the ownership and management groups to enjoy their share of the fruits of increasing productivity. It is very probable that the profit record of the industry would have been substantially worse, however, had the productivity increase not occurred.

The gains from the increased labor productivity in the tire industry, as well as those from other economies, have been widely distributed. The great mass of tire consumers have received the greatest gains and the wage-earners in the industry have received most of the remainder. The industry has operated under conditions of competition with relatively flexible prices, high wages and short hours. Most experts agree that these are the most socially advantageous ways of distributing the gains of increased labor productivity. One result of such a distribution of productivity gains is that the labor displaced by technological change may be most easily reabsorbed into industry. Consumers may use the money saved to buy more tires, thus increasing production and employment in the tire industry, or, they may buy more other products, thereby increasing production and employment in those industries. Likewise, the workers in the tire industry and other groups which share in the gains have the same choice. In the years of the tire industry's most rapid growth, these adjustments caused the production of tires to expand so rapidly that not only were those displaced by the industry reabsorbed, but the employment total grew rapidly. In the last two decades, and especially since 1929, this situation no longer existed. The fact that large numbers of the displaced workmen could not find other jobs is due, not to the increasing productivity of the tire industry, but to the increased frictions and emergency conditions affecting the entire economy.

Although the shifts in the location of the industry are by no means entirely or even largely due to productivity changes, the two movements are closely associated. It has been noted that the rubber industry got its start in the United States about one hundred years ago. It was then located almost exclusively in the New England and Middle Atlantic states. In the next several decades the industry spread out especially over the Middle West. Passing unnoticed at this time was the location

SUMMARY AND CONCLUSIONS 185

of several small companies in and around Akron, Ohio. In the 1890's, when an eastern group of companies almost monopolized the manufacture of rubber footwear and mechanical goods, the principal lines of products of the rubber industry at that time, the Akron companies were eagerly seeking new non-competitive lines of products. Solid rubber carriage tires and pneumatic bicycle tires proved to be one answer to the problem. This put the Akron companies in the best position to manufacture tires for the new automobiles, which were just being introduced on a commercial scale in the late 1890's and early 1900's.

Capitalizing on the enormous expansion in automobile tire demand the Akron companies grew rapidly, and by 1914 Akron was the recognized rubber and especially tire center of the country. In that year Akron companies manufactured more than one-third of all rubber products and a larger proportion of the tire output. Between 1919 and 1929 Akron increased its output to well over 50 percent of the nation's tire production. The Akron factories were the leaders in mass production and scientific management methods, and they became the largest and most efficient tire factories in the world. Although hourly wages averaged from 25 to 30 percent higher than elsewhere in the tire industry, the productivity differential more than compensated for the difference. Even in the period between 1929 and 1935 Akron gained slightly upon the rest of the country, as is indicated by the fact that in 1929 66.4 percent of the total tire wage-earners were in Ohio, while by 1935 the percentage had risen to 68.4.

In the 1920's, however, the beginnings of the decentralization movement were made as branch or affiliated companies were established in several foreign countries. One medium-sized company moved out of Akron in this period and the Akron companies established three branches in Los Angeles and one in Gadsden, Alabama. This movement was more than offset by the growth of the Akron companies prior to 1929. In the lean years between 1929 and 1935 decentralization virtually ceased, but beginning in 1935 a new movement from Akron reduced Ohio's proportion of tire wage-earners from 68.4 percent in 1935 to 58.7 percent in February, 1938. Several of Akron's outside competitors improved their positions substantially, and the Ford Motor Company built an ultra modern tire plant at Dearborn, Michigan, for the purpose of producing a large part and perhaps eventually all of its own requirements. Moreover, the Akron companies themselves expanded the capacity of several of their outside branches, built three new tire producing branches outside of Akron, and shifted the major proportion of their non-tire products to seven new branches outside of Akron. The exodus of non-tire products from Akron became so rapid between 1935 and 1938 that in the latter year only one company was producing any considerable proportion of its non-tire products in Akron. In three years Akron lost nearly as large a proportion of the rubber industry as it had gained during the previous decade and a half of concentration in Akron.

Several factors account for these tendencies. In the first place a certain amount of decentralization (especially that which occurred in the 1920's) has been for the purpose of securing lower transportation costs. Secondly and more important is the fact that the technological developments, which for a long time gave the advantage to the larger units, have shifted in the direction of more individualized equipment since about 1929. Thus it is now possible to operate a modern plant with a capacity of from 5,000 to 10,000 tires per day at costs which compare favorably with those of a plant several times that size, provided the smaller plant specializes on a few types and sizes. The Akron plants have also been slower in plant modernization and the adoption of new processes than the outside plants in recent years. At the very time when the productivity differential of Akron over the outside area has declined, however, wage rates have risen sharply. The Akron companies long paid hourly wage rates of from 25 to 30 percent higher than the rest of the country, but since 1935 they have been paying between 40 and 50 percent more than the outside area (excluding Detroit and Los Angeles). It is to this area that Akron has been losing business. The growth of unionism in the industry has also reacted to the disadvantage of Akron. That city immediately became the center of the industry's labor strife and later emerged as the greatest stronghold of the union. As a result the problems of adjustment to the new conditions of labor relations have been more difficult in Akron than elsewhere in the industry.

The Outlook for Future Increase In Productivity

The productivity of labor is determined by so complicated a set of conditions that any attempt to predict future trends must necessarily be hazardous. Some basis, however, has been laid for an appraisal which may lead to tentative conclusions. First, it must be recognized that in so far as the upward trend of productivity has been attributable to the expansion of the industry, the immediate future presents a different picture from that of ten or fifteen years ago. Automobile registrations are still increasing but the rate of growth has been levelling off. The increase in average tire-mileage and the consequent decline in renewal tire sales per registered car may be expected to prevent renewal tire sales from reaching the levels of the late 1920's. Annual renewal sales averaged nearly 45,000,000 from 1925 to 1929, whereas they have averaged slightly less than 31,000,000 per year between 1934 and 1938. They may be expected to vary between 30,000,000 and 40,000,000 in the next few years. An estimate of the annual tire requirements for original equipment for new automobiles varying between 15,000,000 and 25,000,000 would seem to be liberal enough to allow for some probable expansion in this direction. Other sales are not expected to exceed 3,000,000 or at a maximum 6,000,000 tires per year in the near future. This means that the industry can look forward to an annual output of about 55,000,000 tires in reasonably good years, with a possibility of about 70,000,000 tires under exceptionally favorable circumstances. In serious depression years the output may fall as low as 40,000,000. An annual production of 55,000,000 tires is slightly above the 1936-1937 average and is over one-third more than the 1938 production. A yearly output of 66,000,000 is slightly greater than the 1925-1929 average and represents a greater output than any previous year except 1928 and 1929. On a poundage basis, however, a production of 55,000,000 present-day tires represents more than a 15 percent increase over 1929, and an output of 66,000,000 tires represents nearly a 40 percent increase over 1929. Thus the tire industry may be expected to show a moderate growth within the next five or ten years, but the rate of expansion will undoubtedly be significantly smaller than during the decade preceding 1929. The slower rate of growth will probably be reflected in a similar retardation in the rate of productivity increase. Moreover, mechanization of the main processes has been largely accomplished so that much less than formerly remains to be done in this direction. The same may be said to hold true with regard to motion-time analysis and to the concentration of the industry in larger units.

Nevertheless, it is confidently expected that the next decade will see very considerable advances in labor productivity. It is estimated that if all plants were to be brought up to the standards prevailing in the most efficient factories, the industry would show a productivity increase of at least 20 to 25 percent. Furthermore, the variations between plants are considerable, no one plant being the most efficient in every department. Thus the application of existing techniques could easily result in some increase in the labor productivity of even the best plants of the present day. It is also anticipated that continued advances will be made in technology and in scientific management, resulting in further productivity increases. The flow of new capital into the industry may be expected to continue the progress of mechanization. In spite of the great advances which have already been made, the industry remains less mechanized and utilizes fewer automatic processes and equipment than such mass production industries as steel, glass, or automobiles. It is also probable that substantial increases in productivity can be achieved by the improvement of plant layouts. The newest and most modern plants in the industry are one-story buildings, laid out so as to provide for a better sequence of operations and a practically continuous flow from raw materials to finished product with a minimum of intraplant transportation. The modernization and rearrangement of many of the older plants would result in considerable savings in labor as well as in other expenses.

The continuation of the trend toward higher productivity in the tire industry probably depends more upon the state of labor relations than upon any other single factor. It is a commonplace observation that productivity is in general highest where the morale of the labor force is high and where labor and management enjoy each other's confidence to the fullest extent. It may be that these conditions will be most fully realized when labor and management meet on equal terms without an undue amount of coercion being exercised by either group upon the other. A complete understanding of the large areas in which the interests of labor and management are identical, a feeling of confidence and trust concerning the motives and good faith of each group on the part of the other, and a frank and open discussion of differences appear to be the essentials of the best employer-employee relations.

Employer-employee relations were quite harmonious and a sufficiently high labor morale was maintained to make possible extraordinary gains in productivity in the twenty years prior to 1932. In this period employer-employee relations were controlled by the employers. Independent unions were non-existent. A few companies maintained employee representation systems but they were limited in their range of action. Since 1933, and particularly after 1936, the former pattern of industrial relations has been disrupted by the growth of trade unionism to the point where a substantial proportion of the employees of the industry are now unionists. In the ensuing struggle, violence and coercion were freely employed on both sides and hatreds and animosities were aroused. The confidence between management and labor was seriously undermined and each looked upon the other as an enemy rather than as a partner. The excesses resulting from the struggle for organization have now largely passed, however, and many policies which gave rise to the greatest amount of bitterness on the part of both groups have been modified or eliminated. Several plants which are largely unionized, including the leading producers of the industry, have arrived at a stage of peaceful working relations between the union and the managements through the machinery of collective bargaining. In these plants labor relations are gradually becoming more satisfactory, the bitterness is dying out, and there is a growing feeling of mutual confidence between the parties to the discussions.

There are still many difficulties involved in industrial relations, however. Among the most important are the rights of minorities, union and non-union. The competition between unionized and non-unionized areas having different levels of wages, productivity, labor costs, and standard weekly hours also raises some knotty problems. Likewise the speed of production processes remains a live issue. The unions are opposed to what they regard as undue speeding-up on the grounds of health and safety as well as fear of unemployment. Employers are inclined to discount these arguments heavily. They point out that it is impossible to get capital for plant modernization or new equipment unless they have assurances that both the existing and the new equipment will be used to secure maximum efficiency.

The answer of the United Rubber Workers to some of these problems lies in their attempt to organize the non-union areas and thus to secure more nearly uniform conditions in the industry. Their ultimate goal is to remove the element of wage rates as a basis for competition between plants. This would seem to require, not uniform wage rates throughout the country, but geographical and plant differential adjustments to make labor costs uniform. The success of this program depends not only upon the zeal and efficiency of the union leadership and the record of their achievements in the organized areas, but also upon the attitude of the general public and of the national, state, and local governments. This work promises to be a long and slow development, but regardless of the eventual outcome, it appears that union organization is here to stay for at least the immediate future in Akron, Detroit, and a few other areas. Since the industry is gradually adjusting itself to this situation and the union has shown a recent tendency to be more moderate in its demands, the immediate prospects favor fairly good labor-management relations and a continuation of the slowly rising trend of productivity in the unionized areas. But more permanent and mutually satisfactory agreements between management and labor on the controversial problems of wages, hours, standard work requirements, and the methods and procedures of collective bargaining seem to be the prerequisites of harmonious labor relations in the industry in the long run. In the meantime a certain amount of labor unrest may be expected to continue in the industry.

Other Probable Future Trends

The growth of production of the industry will probably continue at a slower rate than the growth of productivity. The problem of technological displacement of labor is, therefore, likely to remain important. Probably relatively little of this unemployment can be offset by further reduction in the hours of labor. The continued development of new products by the rubber industry will doubtless provide a partial answer to this problem. If the industry should develop synthetic rubber production to a point where it forms an important source of raw materials this would also help a great deal. Synthetic rubber, however, still has a long way to go before it can compete in quality and price with the natural product. Moreover, most synthetic rubber development is being done by the chemical rather than the rubber industries at the present time. Thus the reabsorption of displaced rubber workers into employment will depend in the future, as it has in the last few years, more largely on general business conditions than upon factors within the tire and rubber industries. The way in which the tire industry distributes the gains resulting from increased productivity will, however, remain a significant factor.

The improvement of the quality of the product will accrue largely to consumers, and this group will also get a large share of the other productivity gains. The growth of a vigorous labor movement in the industry, however, indicates that labor may be able to get a larger proportion of the gains in the form of higher wages and perhaps also shorter hours than it has in the past. The financial standing of the industry has been considerably improved since many of the weaker concerns have been either forced out of business or reorganized on a sounder basis. Crude rubber prices have been relatively stable since the advent of the International Rubber Regulation Committee of 1934. The development of synthetic rubber has reached the stage where it can act as a deterrent to a runaway rise in crude rubber prices. The industry is still highly competitive on both a price and quality basis, but the price fluctuations have been more moderate in recent years. All of these factors seem to indicate more stable and higher earnings than the industry has enjoyed in years. Thus it appears likely that ownership and management will be able to secure a somewhat larger share of the gains of productivity increase than they have in the last two decades.

The most difficult factor of all to predict is the future trend of the location of the industry. The situation as of 1939 may be summarized briefly. Akron is still the leading tire center of the United States. More than 35 percent of the national production of tires comes from Akron as well as nearly 10 percent of the other rubber products. The Akron factories have lost their clear-cut productivity advantage over the rest of the nation, but the number of plants which are equal or superior to the Akron producers is small. Akron still has several advantages. Tremendous investments in tire factories are concentrated there, a skilled labor force is available, and the city is close to both the original equipment and the large replacement market centers. Plants and equipment depreciate or become obsolete relatively rapidly, however. A skilled labor supply is less important now than formerly, since the experience of several new plants indicates that the skills necessary for nearly all of the jobs in a tire factory can be developed in a few months. Thus the old plants and equipment will not be replaced in Akron unless the producers there have a favorable cost position.

Whether Akron will continue to lose its share of the industry or not is the chief question at issue, since it is not likely to regain much of the recent losses in the near future. The answer is to be found in the developments in labor costs and labor relations. The 1939 situation under which the Ohio-Michigan area was paying 30 cents per hour or 42 percent higher wages than the outside area is unfavorable to Akron. The Akron factories probably cannot afford to pay wages more than about 25 percent higher than the national average on the basis of their productivity differential. Most of the Akron plants cannot afford this large a differential above the wages of the new branch plants. The same may be said for the wage differentials between the Akron producers and their large and medium-sized competitors located elsewhere. Yet recent events indicate that the United Rubber Workers are determined that any readjustment in the differential must come from wage increases elsewhere rather than from reductions in Akron. The Fair Labor Standards Act of 1938 will not help much because tire manufacturing is a relatively high wage industry. Even the lowest wage plants have few workers earning less than 40 cents per hour, the highest minimum wage which can be set under this law. If the union succeeds in organizing the outside area and in raising wages there, the main impetus to movement of the industry will be removed. Although the results of the union's attempts in this direction are meager to date, the possibility should not be ruled out entirely. Most of these plants are in vigorously anti-union territory, but such sentiment is often subject to relatively rapid change. Akron was an outstandingly anti-union town for decades; yet it was organized in a comparatively short time.

If, however, the union campaign to organize the outside area fails, the wage rates paid in that area will probably not increase sufficiently to narrow the differential between Akron and the outside area. Another alternative is for the Akron plants to increase productivity to a point where the present wage differentials can be maintained without loss. All that is certain is that Akron will continue to lose its share of the tire industry unless either its wage differential over the outside area is reduced or its productivity differential is increased.

A SELECTIVE BIBLIOGRAPHY

Books

Barron, Harry, *Modern Rubber Chemistry*, New York, Chemical Publishing Company of New York, Inc., 1938.

Beney, M. Ada, *Wages, Hours and Employment in the United States, 1914-1936*, New York, National Industrial Conference Board, 1936.

Carlsmith, Leonard E., *The Economic Characteristics of Rubber Tire Production*, New York, Criterion Linotyping and Printing Company, Inc., 1934.

Davis, Carrol C. and Blake, John T. (Editors), *Chemistry and Technology of Rubber*, New York, Reinhold Publishing Corporation, 1937. Published under the auspices of the Rubber Division of the American Chemical Society.

Firestone, Harvey S., and Crowther, Samuel, *Men and Rubber*, New York, Doubleday, Page and Company, 1926.

The Firestone Tire and Rubber Company, *Rubber, Its History and Development*, 1922.

Frazier, C. E. and Doriot, G. F., *Analyzing Our Industries*, New York, McGraw-Hill Book Company, 1932.

Geer, W. C., *The Reign of Rubber*, New York, The Century Company, 1922.

Glover, John G. and Cornell, William B., (Editors), *The Development of American Industries*, New York, Prentice-Hall, 1933. Chapter XII, "The Rubber Industry", A. L. Viles and A. C. Grimley, pp. 227-244.

Goodyear, Charles, *Gum-Elastic and Its Varieties, With a Detailed Account of Its Applications and Uses, and of the Discovery of Vulcanization*, Volumes I and II, New Haven, Conn., printed privately 1853 and 1855. Reprinted by *The India Rubber Journal*, London, England, 1936 and 1937.

Hamilton, Walton and Associates, *Price and Price Policies*, New York, McGraw-Hill Book Company, 1938. Section III "The Automobile Tire —Forms of Marketing in Combat" by Albert Abrahamson, pp. 83-116.

Lawrence, James C., *The World's Struggle With Rubber*, New York, Harper and Brothers, 1931.

Mills, Frederick C., *Economic Tendencies in the United States*, New York, National Bureau of Economic Research, 1932.

——, *Prices in Recession and Recovery*, New York, National Bureau of Economic Research, 1936.

National Industrial Conference Board, *Machinery, Employment and Purchasing Power*, New York, 1935.

Nourse, E. G., *America's Capacity to Consume*, Washington, The Brookings Institution, 1934.

Pearson, Henry Clemens, *Rubber Machinery; An Encyclopedia of Machines Used in Rubber Manufacture*, New York, The India Rubber World, 1915.

Schidrowitz, Philip, *Rubber*, London, Methuen and Company, 1911.

Whittlesey, Charles R., *Governmental Control of Crude Rubber*, Princeton, N. J., Princeton University Press, 1931.
Wolf, Howard and Wolf, Ralph, *Rubber: A Story of Glory and Greed*, New York, Covici-Friede, 1936.

OTHER SOURCES

(Including Articles, Bulletins, Periodicals, Reports, Special Studies, and Unpublished Manuscripts)

Automobile Facts and Figures, 1937 Edition, Automobile Manufacturers Association, Inc., New York.

Barker, P. W., *Rubber: Some Facts on Its History, Production and Manufacture*, United States Bureau of Foreign and Domestic Commerce, Washington, 1936 (Multigraphed).

Beal, Arthur F., "Dispersion in Man-Hour Productivity Since 1929", *Proceedings of the American Statistical Association*, Volume XXIX, March, 1934, Supplement, pp. 66-71.

Beights, David M., *Financing American Rubber Manufacturing Companies*, Unpublished Doctoral Dissertation, University of Illinois, Urbana, Ill., 1932. Also Published Abstract of the Dissertation.

Bliss, Charles A., *Production in Depression and Recovery*, National Bureau of Economic Research, Bulletin No. 58, New York, November 15, 1935.

——, *Recent Changes in Production*, National Bureau of Economic Research, Bulletin No. 51, New York, January 28, 1934.

Bowden, Witt, "Labor in Depression and Recovery, 1929 to 1937", *Monthly Labor Review*, Volume XLV, November, 1937, pp. 1-37.

Burkman, Eric C., "The United States Rubber Company" *Encyclopaedia Britannica*, Fourteenth Edition, 1929, Volume XXII, p. 852.

Cross, W. H., *The Rubber Manufacturing Industry*, National Recovery Administration, Division of Research and Planning, Evidence Study No. 35, Washington, 1935 (Mimeographed).

Cross, W. H., Earseman, G. S., and Lenaerts, J. H., *The Rubber Industry Study*, National Recovery Administration, Division of Review, Work Materials No. 41, Washington, 1936 (Mimeographed).

Croxton, Fred C., Croxton, Frederick E., and Croxton, Frank C., *Average Annual Wage and Salary Payments in Ohio*, United States Bureau of Labor Statistics, Bulletin 613, Washington, Government Printing Office, 1935.

Croxton, Fred C., Lapp, John A., and Hanna, Hugh S., *Findings and Recommendations of the Fact Finding Board Appointed by the Secretary of Labor, November 15, 1935*, Wage and Hour Controversy at the Goodyear Tire and Rubber Company Plants at Akron, Ohio. Washington, Dec. 16, 1935 (Mimeographed).

Dalrymple, Sherman H., "The United Rubber Workers of America", *Labor Information Bulletin*, Volume VI, No. 4, United States Bureau of Labor Statistics, Washington, Government Printing Office, pp. 4-7.

BIBLIOGRAPHY

"Digest of Material on Technological Changes, Productivity of Labor and Labor Displacement", *Monthly Labor Review,* Volume XXXV, November, 1932, pp. 1-27.

Drucker, Mary J., *The Rubber Industry in Ohio,* National Youth Administration in Ohio, Occupational Study No. 1, Columbus, Ohio, December, 1937 (Mimeographed).

Federal Trade Commission, *In the Matter of the Goodyear Tire and Rubber Company,* Docket 2116, Washington, Government Printing Office, 1936.

Fidler, A. T., "Dunlop Rubber Company, Ltd.," *Encyclopaedia Britannica,* Fourteenth Edition, 1929, Volume VII, p. 743.

Firestone, Harvey S., Jr., *The Romance and Drama of The Rubber Industry,* Akron, Ohio, Firestone Tire and Rubber Company, 1932.

Fluctuation in Employment in Ohio 1914-1929, United States Bureau of Labor Statistics, Bulletin No. 553, Washington, Government Printing Office, 1932.

Handbook of Labor Statistics, 1936 Edition, United States Bureau of Labor Statistics, Bulletin No. 616, Washington, Government Printing Office.

Handbook of Labor Statistics, 1929 Edition, United States Bureau of Labor Statistics, Bulletin No. 491, Washington, Government Printing Office.

Hinrichs, A. F., *Akron Rubber Report,* Memorandum to John R. Steelman, Director of the United States Conciliation Service by A. F. Hinrichs, Chief Economist of the United States Bureau of Labor Statistics, Washington, April 27, 1938 (Unpublished Manuscript). Summaries of the report appeared in the *Akron Beacon-Journal* and the *Akron Times-Press* on May 14, 1938.

Holt, E. G., *International Shipments of Automobile Casings, Rubber Boots and Shoes, Rubber and Balata Belting, and Rubber Hose and Tubing from Principal Manufacturing Countries,* United States Bureau of Foreign and Domestic Commerce, Circular Ru-3544, Washington, 1934 (Mimeographed).

——, *Rubber News Letter,* June 15, 1937, United States Bureau of Foreign and Domestic Commerce, Leather and Rubber Division, Circular No. 3644, Washington (Mimeographed).

——, *United States Renewal Tire Market Analysis,* United States Bureau of Foreign and Domestic Commerce, Rubber Section, Special Circular No. 3500, Dec. 1, 1933, Washington (Mimeographed).

The India Rubber Journal, Quarter Century Number, 1909, London, Maclaren and Sons, Ltd.

Jackson, James S., "Decentralization in the Tire Industry", *The Akron Beacon-Journal,* Akron, Ohio, November 16 to December 6, 1937. Series of 16 Articles.

"John Boyd Dunlop," *Encyclopaedia Britannica,* Fourteenth Edition, 1929, Volume VII, p. 743.

Johnson, Elizabeth A., *A Selected List of the Publications of the Bureau of Labor Statistics,* 1936 Edition, United States Bureau of Labor Statistics, Bulletin No. 624, Washington, Government Printing Office.

BIBLIOGRAPHY

Kettering, Charles Franklin, "Motor Car" *Encyclopaedia Britannica*, Fourteenth Edition, 1929, Volume XV, pp. 880-897.

Kress, A. L. and Pearce, C. A., *Material Bearing on the Rubber Tire Industry*, National Recovery Administration, Division of Research and Planning, Washington, 1933 (Mimeographed).

Lederer, Emil, *Technical Progress and Unemployment*, International Labor Office, Studies and Reports, Series C., No. 22, Geneva, 1938.

Leigh, W. W., "1938 Prospects for the Independent Tire Dealer," *Tire Review*, January, 1938, pp. 9-11.

——, "30,000,000 Tires Go to Market", *Tire Review*, June, 1938, pp. 12-15.

——, "The Wheels of Fortune", *Tire Review*, February, 1938, pp. 10-18.

Magdoff, Harry, Siegel, Irving H., and Davis, Milton B., *Production, Employment and Productivity in 59 Manufacturing Industries*, National Research Project of The Works Progress Administration, Report No. S-1, Parts I, II and III, Philadelphia, 1939.

Martin, G., "Rubber: Botany, Cultivation and Chemistry", *Encyclopaedia Britannica*, Fourteenth Edition, 1929, Volume XIX, pp. 602-605.

Mills, Frederick C., "Industrial Productivity and Prices", *Journal of the American Statistical Association*, June, 1937, Volume XXXII, pp. 247-262.

Osberg, E. V., "The New Ford Tire Plant", *India Rubber World*, June 1, 1938, pp. 53-64.

Paull, Wallace H., "Tyre", *Encyclopaedia Britannica*, Fourteenth Edition, 1929, Volume XXII, pp. 653-656.

Peterson, Florence, *Industrial Relations in 1938*, United States Bureau of Labor Statistics, Serial No. R. 905, Washington, Government Printing Office, 1939.

——, *Strikes in the United States, 1880-1936*, United States Bureau of Labor Statistics, Bulletin No. 651, August, 1937, Washington, Government Printing Office.

Proceedings of the First Constitutional Convention of the United Rubber Workers of America, Affiliated with the American Federation of Labor, Akron, Ohio, September 12-17, 1935.

Proceedings of the First Convention of the United Rubber Workers of America, Affiliated with the Committee for Industrial Organization, Akron, Ohio, September 13-21, 1936.

Proceedings of the Second Convention of the United Rubber Workers of America, Affiliated with the Committee for Industrial Organization, Akron, Ohio, September 12-20, 1937.

"Productivity of Labor in Eleven Manufacturing Industries", *Monthly Labor Review*, Volume XXX, March, 1930, pp. 1-17.

"Productivity of Labor in the Rubber Tire and Iron and Steel (Revised) Industries", *Monthly Labor Review*, Volume XXIII, December, 1926, pp. 28-34.

Raymond, Albert, Paper Presented to the American Institute of Mining and Metallurgical Engineers, State College, Pennsylvania, October 20, 1934 (Mimeographed).

BIBLIOGRAPHY

"Rates of Wages, Fluctuation of Employment, Wage and Salary Payments in Ohio", 1923, Bulletin of the Department of Industrial Relations, and the Industrial Commission of Ohio, Division of Labor Statistics, Report No. 8, Columbus, Ohio, 1924, F. J. Heer Printing Co.

——, 1928, Report No. 19.

——, 1929, Report No. 26.

Reynolds, Lloyd G., "Competition in the Rubber Tire Industry", *American Economic Review*, Vol. XXVIII, No. 3, Sept. 1938, pp. 449-468.

Scudder, Stevens and Clark, Investment Counsel, *Report on the Automobile Tire Industry*, Boston, 1931 (Mimeographed).

Shaw, Jerome T., "Tire Sales Volume Recovers" *The New York Times*, November 13, 1938, Section A, p. 26.

Statistical Bulletin of the International Rubber Regulation Committee, London, Monthly since 1934.

Stern, Boris, *Labor Productivity in the Automobile Tire Industry*, United States Bureau of Labor Statistics, Bulletin No. 585, July, 1933, Washington, Government Printing Office.

Stewart, Estelle M., *Handbook of American Trade Unions*, 1936 Edition, United States Bureau of Labor Statistics, Bulletin No. 618, Washington, Government Printing Office.

A Trip Through Rubberland, Akron, Ohio, The B. F. Goodrich Company, 1936.

"Uncertainties in Tires and Rubber", *Barron's, The National Financial Weekly*, New York, October 28, 1935.

Wages and Hours of Labor in the Automobile Tire Industry, 1923, United States Bureau of Labor Statistics, Bulletin No. 358, April, 1924, Washington, Government Printing Office.

Weintraub, David and Posner, Harold L., *Unemployment and Increasing Productivity*, National Research Project of the Works Progress Administration, Philadelphia, March, 1937. Reprinted from *Technological Trends and Their Social Implications*, National Resources Committee, Washington, 1937, Government Printing Office, pp. 67-87.

Whittlesey, Charles R., "Rubber", *Encyclopaedia of the Social Sciences*, The Macmillan Company, 1934, Volume XIII, pp. 453-461.

INDEX

Abrahamson, Albert, see also Hamilton, Walton and Associates, 32
Accidents, industrial, 66, 67
Affiliated companies, see Branch and affiliated companies
Akron, Ohio, 34, 64-66, 99, 102, 103, 107, 108, 115, 142, 146, 150-157, 160-175, 180, 185-187, 191-194
Allied Metal and Rubber Workers Union, 108
Amalgamated Rubber Workers Union of America, 108
American Federation of Labor, 100, 102, 108, 111-113, 115
Armstrong Rubber Company, 168
Assets, see also Capital and capitalization, 85-87
Atlas Supply Company, 168
Automobile Manufacturers Association, 27, 31
Automobiles, 26-28, 31, 32, 38, 39, 42, 122
 production of, 27, 152
 registrations of, 27, 29, 55, 159, 187

Balloon tires, see also Production of tires, types, 43-45
Beal, Arthur F., 18, 19
Bedaux wage system, 96, 97
Beights, David M., 132, 133
Beney, M. Ada, 63, 141
Blake, John T., 48
Bliss, Charles A., 80
Blowouts, see also Quality of tires, 45, 50
Bowden, Witt, 20, 82
Branch and affiliated companies, domestic, 161-163, 169-171, 185
 foreign, 159, 160, 164, 185
Business cycle, 118-125, 155

Cadillac, Mich., 170
California, 161, 162, 165, 166, 172
Capital and capitalization, 33, 85-87, 89, 120, 126, 132, 133, 154, 155, 158, 179
Carlsmith, Leonard E., 32, 39, 40
Centralization of the tire industry, 34, 64, 152-157
Chickopee Falls, Mass., 165
Clarksville, Ky., 170, 175
Clayton Anti-Trust Law, 56, 57

Collective bargaining, see Unions
Competition in the tire industry, 33, 41, 42, 55-57, 131, 156, 158, 160-172, 184, 193
Concentration in the tire industry, see also Establishments, number of, 72, 153-157, 181, 188
Congress of Industrial Organizations, see also United Rubber Workers of America, 33, 34, 112-114
Cord tires, 44, 45
Costs of the tire industry,
 labor, see also Wages, 85, 100, 101, 117, 129, 130, 155, 163, 170, 174, 189, 193, 194
 other, 12, 127, 130, 134, 153, 155, 158, 161, 166, 183, 186, 189
Cotton fabric, see Tire fabric
Crider, John W., 57
Cross, W. H., 33, 39, 46, 57, 60, 85, 134, 139, 159
Crowther, Samuel, 25, 28
Croxton, Fred C., 66, 98-103, 106, 143
Cumberland, Md., 160, 161, 165, 169

Dalrymple, Sherman H., 103, 111, 113, 115
Davis, Carrol C., 48
Dearborn, Mich., 92, 169, 186
Decentralization, 23, 117, 155, 157-175, 185-187
 reasons for, 171-175, 186, 187
Detroit, Mich., 25, 99, 102, 103, 107, 115, 146, 147, 153, 158, 162, 165, 167, 168, 172, 191
Doriot, G. F., 49, 64
Drucker, Mary J., 67, 91, 105
Dunlop, John Boyd, 26
Dunlop Rubber Company, 26, 160

Earseman, G. S., 46, 57, 60, 85, 134, 139, 159
Eau Claire, Wis., 158
Economic characteristics of the tire industry, 30-34
Employment, see also Unemployment, 21, 33, 62-67, 99, 101, 102, 109, 115, 116, 122-125, 153, 156, 157, 160, 167, 169, 170, 174, 176, 181, 182
Establishments, number of, 30, 87-89, 156-171
Exports of tires, 53, 54, 159, 164

INDEX

Fair Labor Standards Act, 193
Fall River, Mass., 170
Fidler, A. T., 26
Firestone, Harvey S., 25, 28
Firestone Tire and Rubber Company, 28, 56, 114, 135, 136, 151, 152, 155, 158, 162, 164, 169, 170
Fisk Rubber Corporation, 155, 165, 168
Flying Squadron, 94
Ford Motor Company, 27, 92, 118, 169
Frazier, C. E., 49, 64

G. and J. Clincher Tire Association, 28
Gadsden, Ala., 162, 163, 165, 169
Geer, W. C., 25, 27
General Motors Corporation, 27
General Tire and Rubber Company, 136, 154, 165, 170
Geographical distribution of the tire industry, 149-175, 184-187, 192-194
Goodrich, Benjamin F., 150
Goodrich, Tew and Company, 150
Goodrich Company, The B. F., 64, 103, 150, 152, 162, 164, 169-171, 173-175
Goodyear, Charles, 25, 48, 149, 151
Goodyear Tire and Rubber Company, 45, 55-57, 65, 94, 98-102, 107, 113, 135, 151, 152, 155, 161-165, 169, 170
Graham, T. G., 103, 168, 171, 173, 174

Hamilton, Walton, and Associates, 32, 57
Hanna, Hugh S., 66, 98-103, 106, 143
Hinrichs, A. F., 142, 146, 147, 156, 162, 164, 166, 167, 170, 172, 174
Holt, E. G., 36, 38, 39, 42, 43, 53, 54, 87, 139, 159, 160
Hood Rubber Company, 170
Hours of labor, 63, 64, 66, 70, 97-103, 113, 114, 120, 125, 143, 147, 148, 178, 181, 182

Immigration, see Labor supply
Imports of tires, 53, 160
Incentives, see also Wages, 95-97
India Rubber Company, 151
India Tire and Rubber Company, 100, 151
Indianapolis, Ind., 158
Industrial management, see also Scientific management, 94-97
Industrial relations, see also Unions, 106-118, 136, 167, 168, 172-175, 180, 181, 187, 189-191, 193, 194
Industrial Workers of the World, 108
International Association of Machinists, 102
International Rubber Regulation Committee, 58, 192
Inventories of tires, 54, 57, 58

Jackson, James S., 161, 163, 170
Jackson, Mich., 169
Johnson, Elizabeth A., 17

Kelly-Springfield Tire Company, 151, 155, 160, 161, 165
Kettering, Charles F., 27
Kress, A. L., 33, 66, 73, 88

Labor supply, see also Employment, 62-67, 101, 102, 153
  child, 66
  immigration, 65
  scarcity of, 64, 65, 109
  women, 65, 66, 142, 145, 146, 170
Lapp, John A., 66, 98-103, 106, 143
Leigh, W. W., 32, 55, 57
Lenaerts, J. H., 46, 57, 60, 85, 134, 139, 159
Lewis, John L., 113
Litchfield, Paul W., 98, 99, 101, 163
Lorentz, J. E., 100
Los Angeles, Calif., 64, 107, 162, 165, 172, 185, 187
Lytle, Charles W., 97

Markets for tires, 31, 32, 50-58, 168, 179, 187, 188
Marks, Arthur H., 47, 48
Mechanical Rubber Company, 151
Mechanization, see also Technological change, 89-94, 104, 105, 120, 155, 156, 158, 164, 167, 179, 180, 186, 188, 189
Memphis, Tenn., 169
Michelin et Cie., 160
Michigan, 146, 147, 165, 166, 169, 170, 172, 186, 193
Miller, James P., 154, 175
Miller Rubber Company, 150
Mills, Frederick C., 80, 128
Montgomery Ward and Company, 168
Morris, J. P., 150
Motion and time study, 95, 105, 106, 109, 156, 180, 188

Natchez, Miss., 168
National Recovery Administration, 33, 46, 57, 60, 66, 73, 85, 99, 102, 111, 112, 116, 164
New Haven, Conn., 149, 168
New Jersey, 149, 151

INDEX

Noblesville, Ind., 170
Nourse, E. G., 60

Oaks, Pa., 169
Oenslager, George, 48
Ohio, 34, 65, 67, 115, 146, 147, 150, 153, 156, 157, 165-167, 172, 186, 193
Osberg, E. V., 92, 169
Outside area, 146, 147, 166-172, 185-187, 193, 194

Paull, Wallace H., 26
Pearce, C. A., 33, 66, 73, 88
Personnel departments, see Industrial relations
Peterson, Florence, 109, 111, 113
Posner, Harold L., 79, 80
Prices, see Tire prices and Rubber prices
Production of tires, 19, 21, 29, 32, 35-61, 121, 152, 153, 176, 186-188
  methods, see also Technological change, 32, 33, 47-50, 74-76, 89-95, 188, 189
  physical capacity for, 58-61, 86, 163, 169-171, 186
  sizes, 46, 47, 130
  types, 25-31, 42-46, 52, 185
  value, 29-31, 50-52, 130, 152, 153
Productivity in the tire industry,
  effects of changes in, 122-148, 181-187
  factors influencing, 84-121, 178-181
  methods of measuring, 16-24, 68-83, 176-178
  outlook for future increase in, 187-191
Productivity in other industries, 15, 17, 77-83
Profits, 33, 121, 131-137, 154, 158, 160, 183, 184, 192

Quality of tires, 37-50, 59, 84, 93, 130, 131, 179, 192

Retreading of tires, 32, 49, 50
Reynolds, Lloyd G., 135, 136
Roads, improvements in, 38, 39
Robinson-Patman Act, 56
Roxbury India Rubber Company, 149
Rubber chemistry, 39, 47-49, 93, 94, 191
Rubber, crude,
  consumption of, 36-38, 70, 71, 163, 164
  prices of, 38, 47-49, 130, 134, 164, 192

  quality of, 49, 93
  sources of, 25, 49, 158, 161
Rubber products other than tires, 25, 51-53, 138, 149-152, 170, 175, 185, 191
Rubber Goods Manufacturing Company, 151
Rubber Manufacturers Association, 19, 33, 36, 39, 58, 60, 85
Rubber, reclaimed, 38, 47, 48
Rubber, synthetic, 191, 192

Salaried employees, 136-138
Sampson Tire and Rubber Company, 162
Scientific management, 94-97, 104, 106, 107, 110, 155, 180, 181, 188, 189
Scudder, Stevens, and Clark, Investment Counselors, 39, 49, 88
Sears, Roebuck and Company, 55-57, 134, 135, 168
Seiberling Brothers, 151
Selection of workers, 64-67, 103, 104, 120, 121
Seniority rules, 113, 114, 120
Shaw, Jerome T., 131
Sherbondy Rubber Company, 150
Skill, 64, 91, 97, 101, 103-106, 153, 193
Smith, Leonard, 60
Stern, Boris, 19-21, 39, 72-76, 90-92, 97, 106
Stewart, Estelle M., 111
Stevenson Restriction Act, see also Rubber, crude, 38, 45, 49
Strikes, see also Industrial relations, 108, 109, 111-116, 173

Task standardization, 94-97, 139
Taylor, Frederick W., 94, 96
Technological change, see also Mechanization and Unemployment, 16, 17, 89-121, 123-128, 145, 171, 173, 184
Tew, Harvey W., 150
Textile mills, 162, 163
Thomson, Robert W., 26
Tillinghast, Pardon W., 28
Tire,
  building, 74-76, 90, 91, 105
  demand, 31, 32, 50-58, 61, 127, 130, 164, 185, 187, 188
  durability, see mileage
  fabric, 44, 130
  mileage, 38-50, 130, 131
  plies, 45, 46, 74
  prices, 33, 50-52, 56, 127-131, 164, 183

sales, see demand
supply, see Production of tires
weight, 38-40, 59
Training of workers, 104, 105, 180
Trenton, N. J., 108

Unemployment, see also Technological change, 17, 100-103, 109, 110, 116, 120-125, 167, 174, 182
Unions, 33, 100-103, 106-118, 136, 172, 173, 180, 181, 187, 190, 193, 194
United Rubber Workers of America, 34, 102, 103, 105, 111-118, 172, 173, 190, 193, 194
United States Rubber Company, 60, 91, 99, 103, 115, 151-159, 162, 165, 167, 168

Vulcanization, 25, 47, 48, 75, 149

Wabash, Ind., 170
Wages, 33, 66, 95-103, 108, 116, 117, 121, 126, 129, 138-147, 156, 161, 163, 165, 166, 172, 174, 175, 180, 182, 183, 186, 187, 190, 193, 194
Watertown, Conn., 170
Weintraub, David, 79, 80
Western Auto Supply Company, 168
Whittlesey, Charles R., 25, 49
Windsor, Vt., 170
Wolf, Howard and Ralph, 38, 47, 48, 50, 108, 114, 150, 151, 152, 169
Working conditions, see also Accidents and Industrial relations, 66, 94, 95, 104, 105

Bei Fragen zur Produktsicherheit wenden Sie sich bitte an:
If you have any questions regarding product safety,
please contact:

Walter de Gruyter GmbH
Genthiner Straße 13
10785 Berlin
productsafety@degruyterbrill.com